NATURAL RESOURCES OF SOUTH-EAST ASIA
General Editor: OOI JIN BEE

THE SUPPLY OF PETROLEUM RESERVES IN SOUTH-EAST ASIA

THE SUPPLY OF PETROLEUM RESERVES IN SOUTH-EAST ASIA
Economic Implications of Evolving Property Rights Arrangements

CORAZÓN MORALES SIDDAYAO

Issued under the auspices of
the Institute of Southeast Asian Studies, Singapore,
and the East-West Resource Systems Institute,
The East-West Center, U.S.A.

KUALA LUMPUR
OXFORD UNIVERSITY PRESS
OXFORD   NEW YORK   MELBOURNE
1980

Oxford University Press
OXFORD   LONDON   GLASGOW
NEW YORK   TORONTO   MELBOURNE   WELLINGTON
KUALA LUMPUR   SINGAPORE   HONG KONG   TOKYO
DELHI   BOMBAY   CALCUTTA   MADRAS   KARACHI
NAIROBI   DAR ES SALAAM   CAPE TOWN

ISBN 0 19 580466 X

Printed in Singapore by Kyodo-Shing Loong Printing Industries (Pte) Ltd.
Published by Oxford University Press, 3, Jalan 13/3,
Petaling Jaya, Selangor, Malaysia

A.M.D.G.

To Mother

Acknowledgements

THE main conceptual framework for this study was formulated in 1975, but it was not until early 1978 that circumstances allowed me to begin serious research on the subject. Both before and after that starting point, many persons contributed to the work in its present form. Although I will not be able to mention all of them, I would like to identify specific persons to whom I feel deeply indebted. Some were extremely helpful in providing me access to data. Others were helpful in sharing their insights and knowledge of the industry with me. Others provided intellectual stimulation and shared their curiosity as well as their professional expertise. Some, through their very generous hospitality, made my visits to their cities more comfortable as well as more productive than they might have been. Still others were all or most of these. Of special mention then are the following:

Indonesia: Dr. Peter Weldon, Ford Foundation (then in Jakarta and at this writing in Bangkok); Professor Amado A. Castro (ASEAN Secretariat); Professor Dr. Koentjaraningrat (University of Indonesia); Professor Dr. Ir. M. Sadli (former Minister of Mines); Dr. J. Panglaykim and Clara Joewono (Center for Strategic and International Studies); S. L. Rieb (Chief Geologist, Conoco, and a long-time friend); Ir. Wijarso (Director-General of Oil and Natural Gas, Department of Mining and Energy); Ir. E. E. Hantoro (Head, Exploration/Production, Directorate General for Petroleum and Natural Gas); Miss Ann Soekatrie Sabardiman Sosrokoesoemo S. H. (Assistant Chief, Legal Division, MIGAS) and her colleagues, Drs. Sugeng Wibowo, A. Iskandar S. H., W.

Bapangsamirono S. H.; Z. Goeritno S. H., and B. Soeminto S. H. (Pertamina, Legal); Ir. Trisulo (Pertamina Director for Exploration/Production); Ir. Soedarno Martosewojo (Pertamina Director for General Affairs); Mr. Gozali (Secretary, Board of Government's Commissioners, Pertamina); Dr. A. Arismunandar (Director, Electric Power Research Centre); I. B. Gold (Geologist, AMOSEAS); Dr. Thee Kian Wie (National Institute of Economic and Social Research—LEKNAS-LIPI); and Lawrence P. Taylor (Petroleum Officer, American Embassy).

Malaysia: Thomas G. Carson (Senior Geophysicist, Exxon, and an old-time friend); Dr. Goh Cheng Teik (then Deputy Minister for Works and Utilities); Encik Rastam Hadi (Managing Director, Petronas); R. T. Fetters (then Acting Exploration Manager, Exxon); and Mr. Chung Sooi Keong (Director-General, Geological Survey of Malaysia).

Philippines: Apollo P. Madrid (Exploration Manager, Bureau of Energy Development, and an old-time friend); Atty. Wenceslao de la Paz (Director, Bureau of Energy Development); and Lydia R. Vicente (Public Relations Officer, Petrophil, and an old-time friend).

Singapore: Allen G. Hatley (an old-time friend and now Manager of Acquisitions, International Division, Energy Resources Group, Cities Service, in Houston); Christopher S. Kenyon (Geologist, Cities Service); Peter B. Stilley (Joint Venture Coordinator, Union Texas Asia); Salvador de Luna (Geologist, Bow Valley, and an old-time friend); Jean MacDonald (Director, Offshore South-East Asia); Dr. Khin Aung (Geologist, Consultant); Dr. Anne Booth and Dr. Chia Siow Yue (both of the Department of Economics, University of Singapore).

Thailand: Dr. John Ringis (CCOP); J. A. Callow (UN, ESCAP); Prakal Oudom-Ugsorn (Geologist, Natural Gas Organization of Thailand); Preecha Supalak (Mineral Fuels Division, Department of Mineral Resources); Dr. Vatchareeya Thosanguan (Faculty of Economics, Thammasat University); and Vinitha Thummanond (CCOP).

United States: Dr. Richard P. Sheldon (Senior Research Geologist on leave from the U.S. Geological Survey and Coordinator, Raw Materials Systems, East-West Resource Systems Institute); Richard Meyer (Chief, Office of Resource Analysis, U.S. Geological Survey); Dr. Bernardo F. Grossling (U.S. Geological Survey); Lucio D'Andrea, Virginia Yates, and Julia F. Hutchins (all at the Department of Energy); and Professor Moheb Ghali (Department of Economics, University of Hawaii-Manoa).

I would also like to thank the Asia Foundation for the grant that provided funds for the literature research and travel to Bangkok, Jakarta, Kuala Lumpur, and Manila.

Thanks are also appropriate to the Institute of Southeast Asian Studies and the East-West Resource Systems Institute (The East-West Center) for providing the research base and support staff for this study. In these two places it would be amiss for me not to single out certain individuals. First, I would like to thank, in a very special way, Dr. Sharon Siddique for taking over some of my 'chores' during the last four months of my stay at ISEAS; a very good friend and true professional, she thus allowed me to concentrate on the usually difficult phase of writing the first draft. I want to express my deepest appreciation to Dr. Harrison Brown and Dr. Kirk R. Smith for allowing me to do the revisions and additional analytical work in this study, including provision of a research aide. I want to thank the library and administrative support staff at ISEAS, but especially Karthiani d/o Mani Nair for her serious and intelligent secretarial assistance. My special thanks also go to the support staff at EWRSI and at Media Production Services; in an extra special way I thank Thammanun Pongsrikul who so patiently and cheerfully put up with the grubby statistical and computer work related to Chapter VI, and Beverly Takata for her truly efficient secretarial assistance.

My very special appreciation for the moral and intellectual support of Richard A. Bray and of my brothers, Crescencio and Antonio.

To all, including those not mentioned, my sincere thanks and eternal indebtedness.

Needless to say, the responsibility for shortcomings and errors remain with me. The opinions expressed in this study are solely those of the author and do not, in any way, necessarily represent those of the institutes sponsoring the research.

The East-West Center,   Corazón Morales Siddayao
Honolulu, Hawaii,
18 April 1979

Contents

Tables

Figures

Abbreviations

AAPG	American Association of Petroleum Geologists
ADB	Asian Development Bank
AGIP	Azienda Generale Italiana Petroli
Amoseas	American Overseas Petroleum, Ltd.
API	American Petroleum Institute
ARCO	Atlantic Richfield Company
ASCOPE	ASEAN Council on Petroleum
ASEAN	Association of South-East Asian Nations
AWSJ	*Asian Wall Street Journal*
Bbl.	Barrels
B/D	Barrels per day
BP	British Petroleum
Caltex	California-Texas Oil Corporation
CCOP	Committee for the Coordination of Joint Prospecting for Minerals in Asian Offshore Areas
C.F., cu. ft.	Cubic feet
Conoco	Continental Oil Company
ECAFE	Economic Commission for Asia and the Far East
ENI	Ente Nazionale Idrocarburi
ESCAP	Economic and Social Commission for Asia and the Pacific
FEER	*Far Eastern Economic Review*
GDP	Gross Domestic Product
GNP	Gross National Product
GOI	Government of Indonesia
IIAPCO	Independent Indonesian American Petroleum Company
IIASA	International Institute for Applied Systems Analysis

INCA	Indonesia Consortium Activities
IRS	Internal Revenue Service (United States)
JAPEX	Japan Petroleum Exploration Co. Ltd.
LNG	Liquefied natural gas
MBD, MB/D	Thousand barrels per day
MER	Maximum efficient rate (of recovery)
MIGAS	Direktorat Jendral Minyak dan Gas Bumi
MMB	Million barrels
MMCFD	Million cubic feet per day
MNC	Multinational corporation
MRDS	Mineral Resource Development Series (United Nations)
MTCE	Metric tons of coal equivalent
OCS	Outer continental shelf
OGJ	*Oil and Gas Journal*
OPEC	Organization of Petroleum Exporting Countries
P.D.	Presidential Decree (Philippines)
PE	*Petroleum Economist*
Permina	Perusahaan Minjak Nasional
Pertamina	Perusahaan Pertambangan Minyak dan Gas Bumi Negara
Petronas	Petroliam Nasional Berhad
PN	*Petroleum News S.E.A.*
PSC	Production-sharing contract
REDCO	Rehabilitation Engineering Development Company
S.E.A.	Southeast Asia
SEAPEX	Southeast Asia Petroleum Exploration Society
ST	*Straits Times*
Stanvac	Standard-Vacuum Oil Company
Topco	Texas Overseas Petroleum Company
UN	United Nations
UNDP	United Nations Development Programme
UNITAR	United Nations Institute for Training and Research
USBM	United States Bureau of Mines
WAES	Workshop on Alternative Energy Strategies
WEC	World Energy Conference

I

Introduction

SOUTH-EAST ASIA has one of the world's more favourable accumulations of ultimately recoverable petroleum.[1] Each of the host governments in the region has its own goals with regard to the development of these resources,[2] with such goals determined by individual country socio-economic needs.

In the meantime, structural changes have occurred in the institutional framework within which such resource development activities are conducted, both world-wide and in the region. Although for both economic and technical reasons the economic agent directly engaged in the exploration for and development of petroleum in South-East Asia is the private firm, in the last decade the role of the state[3] in determining the level and direction of such activity has become more important than it has ever been.

The allocation of human or capital resources to any economic activity can be (1) determined by a governing body, or (2) determined in the market-place by normal reward incentives. In a mixed society where private capital is involved in furthering that economic activity, the most important motivating factor for drawing that capital is the expectation of a satisfactory return on investment. The underlying factors are (1) a favourable cost/price relationship, and (2) the expectation that the profit margin will continue and justify taking the risk in invested capital. Such profitability may be determined by any of several institutional variables. Although certain technical costs—and the commodity's price itself—may be beyond the control of either the host government or the firm, other elements that enter the cost structure—such as taxes or contractual terms—are within the control of the host government. So are policies that regulate the manner in which the firm disposes of the cost-price residual.

Since it can be assumed that most governments in the region would like to develop their petroleum resources, we can also assume that there is a demand for petroleum reserves; i.e., there is a demand for discovered petroleum resources that can be economically recovered under existing conditions. Since it also can be assumed that the host countries in the region must largely rely on the private firm, generally foreign, to explore for and develop these resources, we can further assume that the supply of the desired petroleum reserves will be forthcoming only if the cost-reward structure attracts the required capital funds for that activity.

BACKGROUND

Before proceeding further, it will be useful to review briefly some basic characteristics of the industry and the relevant conceptual framework in returns maximization. The production of petroleum resources involves three phases: *exploration* (the location of producible reserves), *development* (delineation of reserve volume in a given field), and *production* (actual lifting of the resources from the ground).

Decisions to make the necessary investment are essentially made separately in the first two stages. The decision to spend and search for unknown but producible reserves is predicated on the assumption that both the institutional framework and technology allow profitable production and marketing of the output. The decision to develop producing capability depends on the size of the discovery made in the first stage, assuming the institutional factors affecting the initial decision to engage in exploratory drilling remain unchanged.

A crucial difference between petroleum production and other types of industrial ventures is the degree of risk and uncertainty involved in the first two stages. While the basic concepts of conventional economic decisions may apply—that is, one may assume that under given conditions the producer still maximizes his revenues where long-term marginal revenue equals long-term marginal costs—the approach to arriving at that equilibrium differs.

Basically, the risk considerations are technical in nature. When a firm operates in a foreign country, other uncertainties and risks—

which may be termed 'political. risks'—may be important determinants of investment. After taking such risks into consideration in
the exploration and development stage, a firm computing the present value of its probable income stream must consider several other
factors. In addition to the current rate of production, it must consider these: (1) engineering limits to the rate of extraction in any
given period, (2) physical limits to the total amount of the resource
that can be produced within a given location, and (3) limits to the
availability of new petroleum sources at the same costs as at the
present location.

It might be useful at this point to note that the firm as an exploration agent in South-East Asia operates largely as a contractor to
the host government who has ownership and final control over the
petroleum resources. The group of suppliers of petroleum reserves
in the region is characterized by the presence of the major companies, directly or through subsidiaries, as well as by many smaller, and even independent, companies. Three major companies
currently dominate exploration and producing operations in two
countries—Exxon and Shell in Malaysia, Shell in Brunei. In
Indonesia, Caltex and Stanvac dominate production. Outside
Malaysia and Brunei the bulk of new exploration is conducted by
relatively smaller companies side-by-side with the major companies
like Exxon, Gulf, Mobil, Shell, and Texaco.

FRAMEWORK FOR ANALYSIS

There are several ways of studying the sensitivity of the petroleum firm to changes in the variables of the supply function for
petroleum discoveries or reserves. To a large extent, the approach
will depend on the purpose of the analysis; to a similar extent, the
approach will be defined by the availability of information. In general, studies have not attempted to estimate directly the sensitivity
of the firm's rate of return to investment variables because of the
non-availability of industry data of this nature. Thus, most studies
have tended to concentrate on costs.

Where data are available, an aspect for study could, therefore,
be a test of the sensitivity of wildcat drilling to changes in economic
incentives; that is, to changes in prices and costs (with the latter

including government 'take', the difficulty of the drilling environment, depth, and success ratio, or knowledge of the area).

Another aspect could be a study of the property rights arrangements and the influence of the variables on the petroleum reserves supplied by a firm. For example, a study might address the question of the manner in or degree to which exploration activity is influenced by the framework of taxes, bonuses, production shares, and control over profits (in the form of repatriation or disposition of such profits). It is also possible that the prospect of nationalization might require a firm to consider the need for higher rates of return, which in effect means shorter extraction periods or shorter investment recovery periods.

A third approach would be a study of the industry structure, the pattern of investment in exploration and development, and the resulting level of exploration activity. The following relationships, for example, might be studied; the size of the company in relation to acceptable risk; the relation of the degree of integration to risk; or the amount of front-end commitments, such as bonuses and work expenditures, and its relation to company size.

Some further questions worth investigating relate to factors external to the region. For example, the size of the company's overall operations may affect its behaviour in the region. Factors which affect the company's overall operations but which have nothing to do with host government policies could positively or negatively affect the supply function. Thus, it is possible that external factors (such as technology, or the behaviour of non-regional host governments) rather than internal factors (such as regional demand or regional host government policies) may be responsible for a shift in the supply function.

The present study will focus on the first and second approaches noted above, with thought given in the analyses to the third and fourth. The following main question is addressed in this study: *Is the present system of economic rights, within which exploration agents are expected to operate, achieving the desired (optimal)*[4] *allocation of resources to petroleum resource development in the region?*

This resource allocation question will be studied within the framework of the relationship between property rights[5] and economic decisions. The allocation of resources to exploration and discovery—that is, to the supply of petroleum reserves—will depend on the response by the economic agent to the framework of policies defining the cost-reward structure. Such allocation will be influenced by the net returns that the agent can expect over a well-defined investment period as well as by any intervening policies that may result in any attenuation of such expected returns.

Any firm must operate within the legal, political, social, and economic framework of its sphere of operations. Within these constraints, it must work towards its goal. A business firm, by definition, is profit-oriented, and, therefore, at least in principle, the firm seeks to optimize its operations, i.e., to maximize its preferences (primarily its profits) within its constraints.

On the other hand, the final shape of petroleum legislation in most developing countries, including those of South-East Asia, is determined by the political and socio-economic situation and the resulting goals. The primary target of legislation is to obtain as large positive benefits as is possible for the society from the exploration and development of its petroleum resources, whether the benefits be in the form of foreign exchange earnings or foreign exchange savings, etc. The direct earnings from the operations accrued by the state are a key component of the legislation; these may be in the form of bonuses, surface duties, royalties, corporate income taxes, state participation, output sharing, or various combinations of these methods. The intent is generally to reap these benefits over the long term.

The value to policy analysis of understanding the response to policy variables of the petroleum exploration firm in South-East Asia rests, among other things, on (1) correctly interpreting the goal of the firm, (2) the choice of assumptions underlying any explanation of its behaviour, (3) correctly identifying the explanatory variables and the values attached to them, and (4) the logical consistency of the arguments contained in the analysis. The test of logical consistency is, however, distinct from the empirical truth of the conclusions. As some authors point out, deductive validity does not guarantee empirical truth or even empirical significance.[6] To test a general hypothesis against a specific set of data requires addition of special constraints to a model—and nowhere is that more true than in the study of the petroleum exploration firm in South-East Asia.

An analysis of petroleum exploration in South-East Asia, therefore, cannot be made without reference to the environment in which the exploring firm operates. Rather, it must be made with reference to the firm's relationship not only to others in the industry but to the institutional framework.

The conceptual model in this study thus emphasizes the relationship between firm behaviour and the institutional framework (which term will include the contractual framework, all related policies, and other institutional factors). This differs from the usual approach in the field of industrial organization. Studies usually emphasize the relationship between structure and performance. The following diagram provides an idea of the overall relationship under consideration:

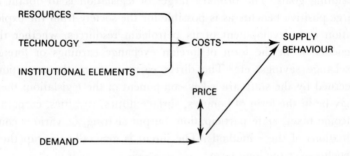

Furthermore, since operationally the supply of petroleum reserves only comes about as a result of wildcat or exploratory drilling, the analysis relates the supply of reserves to the initial investment decisions via the following steps:

1. discoveries are a function of exploratory drilling;
2. exploratory drilling is a function of investment decisions;
3. investment decisions are a function of geological, economic and political factors.

This study recognizes that the involvement of foreign and, often, multinational firms introduces an additional dimension to the decision-making process. The implications of this characteristic will also be considered in the study.

The usefulness of received economic theory in developing the above model may initially give rise to some doubts in the context of South-East Asia.[7] However, the concept of optimization, which is a hallmark of the economist's approach, is sufficiently broad to be able to accommodate several methods of analysis for policy purposes. The property rights approach notes that different property rights arrangements lead to different cost-reward systems, and thereby imply different outcomes of the decision process.[8] If we accept this premise, then an analysis of the interrelation between institutional arrangements and economic behaviour becomes feasible, for the prevailing system of property rights defines the position of the economic agent with respect to the utilization of a scarce resource, the norms of behaviour that must be observed, and the cost of non-observance. In other words, it identifies the basis for optimization by the firm under varying operating conditions.

The model in this study is *dynamic*, that is, the firm has alternative courses of action and possible outcomes which involve more than one time period. The firm's decision variables are subject to certain constraints. The constraints related to the variables of future periods are largely determined by the future environment. The type of *information* the supplying firm has concerning the future may range from one of subjective risk (it has incomplete information for making an objective basis for computing the probability of outcomes) to uncertainty (it is able only to indicate a range of outcomes). Assuming multiple goals that are ranked in order of priority, the firm chooses that alternative action which it considers the 'best' in a given subset of known, possible behaviour alternatives.[9]

DATA AND SCOPE OF THE STUDY

The study covers the nine countries that are traditionally included in the region known as South-East Asia.[10] Some may wonder why I have limited my scope to this region, and why other important countries in the immediate proximity, such as China or Taiwan, were not included. There is really no adequate answer to that question, other than institutional choice and convenience.

Where possible, government-collected data were generally preferred. These were obtained directly from the governments, from government-issued publications, or through the United Nations. In some cases, however, it was necessary—and even preferable—to use industry data. These generally referred to the technical aspects of the study, and this approach was preferred for ease of international comparability.

It might be useful at this point to bring up some of the data problems that were encountered as the study progressed. These problems severely limited the scope of the analyses in some parts. Data on exploration activities have not been collected in a systematic manner throughout the region; thus, data presented in official forums are not necessarily comparable, either among the nations concerned or between regions in the international industry. For example, data on new reserve additions that resulted from revisions of reserve data or extensions in the same structure could be useful tools for long-term analyses of responses to policy changes. The primary problem, to begin with, however, is that reserve data are generally not released by host governments even on an aggregate basis. Similarly, the careful distinction between wildcat discoveries and successful extension tests would be useful, but these distinctions are not always given careful attention in reporting. Data on costs would also be very useful. Understanding the correct level of costs in the region is a key to understanding behaviour of firms in so far as such knowledge allows a proper estimation of probable reaction. These also are not available in a methodically reported form.

Notwithstanding the problems encountered, it is hoped that this study provides a discussion that will be of interest to students of theory and policy. The study is broken down in the following manner:

Chapter II presents an analysis of the resource base within the context of the market in which petroleum is traded. It provides a background discussion of the resource potentials of petroleum and of competing energy goods as well as of the demand pattern for these energy forms.

Chapter III describes the distinguishing characteristics of the petroleum industry, summarizes the factors influencing the levels of exploration and development of petroleum in the region in the post-World War II period, and reviews the structural changes that have taken place in the industry, regionally and globally.

Chapter IV outlines the contractual and legislative framework for petroleum exploration and development in each South-East Asian country. This framework includes the laws, policies, and regulations that determine the property rights structure for the petroleum exploration contractor.

Chapter V analyses the contractual framework and suggests implications for the supply of petroleum reserves of specific aspects of the framework, in the light of the assumptions governing behaviour of firms and of the technical aspects characterizing the industry.

Chapter VI analyses some of the more important exploration indicators, attempts to measure the influence of several variables on exploration drilling, and presents both qualitative conclusions and quantitative inferences. Regression methods were used to estimate the effects of the explanatory variables.

Chapter VII provides the overall conclusions of the study. It suggests further major policy implications of the evolving property rights arrangements, in the context of the resource development goals of the host governments.

1. The term 'petroleum' will be used to refer to both oil and gas in this study, inasmuch as the exploration for petroleum may discover oil or gas, or both.

2. The distinction between 'reserves' and 'resources' is explained in Chapter II.

3. The term 'state' will be used in this study to refer to the central government in any country, regardless of the specific manner in which the political subdivisions of a country are organized in relation to that centre.

4. The term 'optimal' is used to refer to the best value that a variable can take with reference to some particular objective.

5. The term 'property rights' refers to sanctioned behavioural relations among men that arise from the existence and use of economic goods. The relations specify norms of behaviour and the cost of non-observance. See E. G. Furubotn and S. Pejovich, *The Economics of Property Rights* (Cambridge, Mass.: Ballinger, 1974).

6. See K. J. Cohen and R. M. Cyert, *Theory of the Firm*, 2nd ed. (New Delhi: Prentice-Hall, 1974), p. 20.

7. The traditional theory of the firm continuously receives serious criticisms because some of its underlying assumptions fail to correspond with observable or existing institutional structures or conditions. As recently put by Gordon:

'... First, the mainstream of economic theory sacrifices far too much relevance in its insistent pursuit of ever increasing rigor. And, second, we economists pay too little attention to the changing institutional environment that conditions behaviour....'

See Robert A. Gordon, 'Rigor and Relevance in a Changing Institutional Setting' (Presidential address delivered at the 88th meeting of the American Economic Association, Dallas, Texas, 29 December 1975), *American Economic Review*, Vol. 66 (March 1976), p. 1.

8. See E. Furobotn and S. Pejovich, 'Property Rights and Economic Theory: A Survey of Recent Literature', *Journal of Economic Literature*, Vol. 10 (1972), pp. 1137–62.

9. In the neoclassical economic model, also sometimes referred to as the *certainty model*, the firm is assumed to have perfect information about its alternative courses and future events, and can relate a unique outcome to each alternative. Hence, the concept of maximization implies, among other things, that the entrepreneur can choose between several sizes of profits. He knows what his costs are, what his revenues will be, and then makes his decision. Critics have raised questions about the effects of risks and uncertainty on the behaviour of the firm. (See, for example, Frank H. Knight's *Risk, Uncertainty and Profit* (Boston: Houghton Mifflin, 1921) and the Simon-Theil theorem on certainty equivalence in H. A. Simon, 'Dynamic Programming under Uncertainty with a Quadratic Criterion Function', *Econometrica*, Vol. 24 (1956), pp. 74–81, and H. Theil, 'A Note on Certainty Equivalence in Dynamic Planning', *Econometrica*, Vol. 25 (1957), pp. 346–9.) In the absence of perfect information about future events, the firm faces multiple possible outcomes and has a decision problem more difficult than what it faces under the certainty model. As Nordquist points out, however, rationality does not require that a choice is made under conditions where all alternatives and all outcomes are known. (See G. L. Nordquist, 'The Breakup of the Maximization Principle', *Quarterly Review of Economics and Business*, Vol. 5 (Fall 1965), pp. 33–46.) If a decision-maker perceives a set of contemporaneous alternatives, and he is able to rank these in a consistent preference ordering, rationality suggests that he proceeds to choose (maximize) from among that subset of available alternatives that which is most preferred. Rational behaviour thus implies the existence of some maximization procedure, regardless of the state of knowledge underlying the decision.

10. South-East Asia is defined in the conventional sense and covers the following countries: Brunei, Burma, Cambodia (Kampuchea, also previously known as the Khmer Republic), Indonesia, Laos, Malaysia, the Philippines, Singapore, Thailand, and Vietnam.

II

Petroleum and Other Energy Forms: Resource Base and Utilization Rates in South-East Asia

A full appreciation of individual South-East Asian national policies on petroleum resources—that is, policies for obtaining a supply of petroleum reserves—is not possible without understanding the total energy picture in each country. An analysis of the demand for and supply of such reserves requires some explanation of the market in which this resource is traded. The individual characteristics of the resource bases, as well as the patterns of energy consumption, are normally evaluated by policy-makers in terms of each country's socio-economic goals to determine the direction of resource development policies. Thus, a study of the variables that influence exploration for petroleum must begin with a background discussion on the resource potentials of petroleum and of competing energy goods, as well as the demand patterns for these energy forms.[1]

Resource Potentials of the Region

The demand for a resource must be seen in relation to its substitutes—whether these substitutes are perfect or less than perfect in nature. Before proceeding further it is therefore useful to review the alternative energy resources of South-East Asia that may be reasonably tapped within the foreseeable future. In this section, then, the state of knowledge of the bases in the region will be briefly presented.

RESOURCES VERSUS RESERVES

It is pertinent at this point to draw the distinction between the terms 'reserves' and 'resources' as used in this study. The terminology used in international bodies, such as the World Energy Conference, is followed. The term *resources* is used to refer to concentrations of naturally occurring solids, liquids, or gaseous materials in or on the earth's crust; these concentrations may be discovered, undiscovered, or surmised to exist in such form that extraction is currently or potentially feasible. Materials classified as *reserves* are those portions of identified (or discovered) resources that can be produced at a profit and legally extracted at the time of classification. They are, in other words, discovered deposits recoverable with existing technology under current economic conditions. (The term 'ore' is also used to refer to the reserves of some minerals.)

Figure 2.1 presents a scheme showing the relationship between reserves and resources. Resource availability is expressed in terms of (1) degree of certainty based on the extent of geologic knowledge about the existence and characteristics of the resource, and

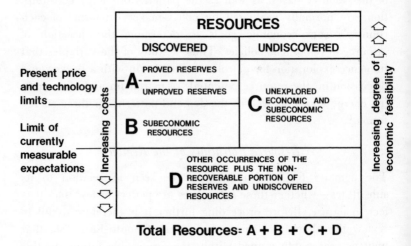

Figure 2.1. Diagrammatic Representation of Mineral Resource Classification

(Modified from McKelvey, 1973, U.S. Geological Survey, and Schanz, 1978, Resources for the Future.)

(2) feasibility of its economic recovery. The degree of certainty is classified into *identified* and *undiscovered*. The feasibility of economic recovery is distinguished by the terms *recoverable, paramarginal*, and *submarginal*. (In discussing a specific resource, more specialized terminology may be desirable. The terminology for petroleum will be presented in the following section. A more complete list also appears in Appendix A.)

The term *identified resources* refers to resources whose location, quality, and quantity are known from geologic evidence supported by appropriate engineering measurements. *Reserves* refer to a subset of identified resources. Resources which have been identified can further be subdivided into *measured*, *indicated*, and *inferred* categories.[2]

Undiscovered resources include those in areas that have not been sufficiently mapped, sampled, or explored but which are surmised to exist, on the basis of current broad geological knowledge and theory. Under this classification fall two others: hypothetical and speculative resources.[3]

The term *recoverable* indicates feasibility of exploitation under current technological and/or economic conditions. The term *economic* is used in the present scheme synonymously with the term *recoverable*.

Following the above taxonomy, identified, measured reserves of petroleum include what the petroleum industry calls *proved reserves* plus what it refers to as 'indicated additional reserves'.

Two methods are generally used in the estimation of potential oil and gas resources. These are (1) the mathematical method, and (2) the volumetric or geologic method.[4] The mathematical method is based on discovery and performance curves; it ignores the geology of the area. The geologic method relates factors that control known occurrences of oil and gas to factors present in prospective areas. It also applies historical data on past exploratory drilling experiences to unexplored, favourable strata. The two methods produce differing results, and a debate concerning the validity and usefulness of the resulting estimates have clouded the issue of future supplies relative to prospective demand. That debate will not be entered into in this study.

RUDIMENTS OF OIL ACCUMULATION AND PRODUCTION

It will also be useful at this point to have a basic understanding of how petroleum is produced.

Fossil fuels—coal, oil, and gas—form in sedimentary basins and geosynclines through the combined action of pressure, temperature and physical-chemical processes on dead marine animal and plant debris over an immense period of time. Large deposits of petroleum require the presence of, or proximity to, thick sedimentary rock strata that were deposited in an appropriate marine environment and suitable geological traps. Chemical change in biological material that was deposited during the last 600 million years in thick layers of sediments (then the earth's surface) produced fossil fuel resources. As other layers of sediment rocks piled over time on top of the original sediment containing hydrocarbons, the pressure forced the petroleum to 'migrate' out of the 'source' rocks until some escaped into the atmosphere or were trapped in the small spaces in porous rock by an impervious layer, or 'cap' rock, with oil often found as a layer between water and gas (see Figure 2.2). Thus the explorer is really not merely looking for petroleum, but for possible petroleum 'traps' where the migrating oil finally accumulated; this 'reservoir' rock is a porous layer (usually sandstone or limestone). Several factors affect reservoir conditions. Some of these are: (1) the ability of the rock formation to hold fluids, or its porosity; (2) the ability of the rock to have fluids transmitted through it, or its permeability, and (3) the pressure in the reservoir.

The presence of a petroleum reservoir can be determined only by drilling into the trap. At the same time, the only actual measurement that can be made is of the oil and gas produced; all other reservoir data are estimated. As a general rule, some gas is always found with oil, but quite frequently gas is also found without oil. Under some reservoir conditions, natural gas is dissolved in the crude oil which, when brought to the surface, releases the gas. Oil in a trap is usually under pressure, and oil production is a displacement process, with gas or water, or both, filling up the portion of the reservoir vacated by oil. This may be done by puncturing the impermeable cap of the reservoir and creating a lower pressure in

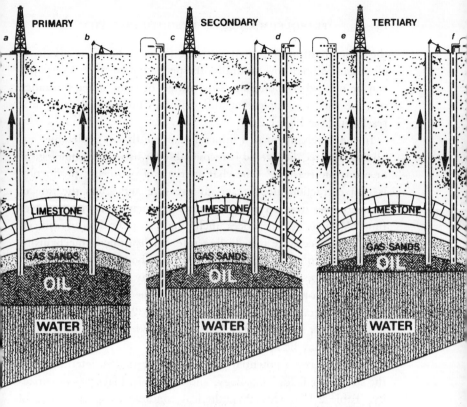

Figure 2.2. Oil Extraction: Primary, Secondary
and Tertiary Recovery Mechanisms

Oil in a trap is usually under pressure. Oil production is a displacement process,
with gas or water, or both, filling up the portion of the reservoir vacated by oil.
Puncturing the impermeable cap of the reservoir creates a lower pressure in the well
bore than in the rock formation. In *primary* production, the oil will flow to the surface (a), or is pumped out (b). If oil will not flow naturally, the necessary differential to cause *secondary recovery* can be achieved by water flooding (c) or gas injection
(d). *Tertiary recovery* methods involve, among other things, the use of steam injection
(e) or chemical injection (f) in high-viscosity reservoirs.

Source: Adapted from Andrew R. Flower, 'World Oil Production', *Scientific American*
    (March 1978), p. 43.

the well bore than exists in the rest of the rock formation. If the
pressure differential is sufficient, the oil or gas will flow through the
well bore to the surface. Gas, an extremely important fuel by itself,
is of critical importance in the production of oil, because it is one of

the major natural or primary driving forces in an oil reservoir. If oil or gas will not flow naturally, the necessary pressure differential can be created by pumping. Gas, water, or some other substance may be used to supplement and maintain the natural drive in a reservoir. (See Figure 2.2.)

In the development of a mineral resource that is depletable, not all the resource discovered is recoverable for use—for technological or economic reasons.[5] One characteristic of petroleum production is that extraction alters reservoir pressure and alters the rate of recovery over time, so that the rate of production may decline over time.[6] Various field practices, including the application of artificial aids, may decelerate this decline rate. Such artificial mechanisms are referred to as 'secondary recovery'[7] methods or 'tertiary recovery'[8] methods. Although the decline rate may be slowed down, not all the 'oil-in-place'[9] may be expected to be recovered. The current U.S. average rate of recovery is around 32 per cent,[10] although in some fields such recovery is as high as 80 per cent, whereas in others this may be as low as 5 per cent.[11]

Each field has a potential production rate that is determined by the size of the field, its geology, and its installed facilities or capacity. Table 2.1 provides hypothetical conditions for three sizes of reservoirs. The initial peak capacity for a small reservoir is shown as about 15,000 barrels per day, with the field staying at this peak capacity for only about one year. In addition, the reservoirs are characterized by depletion periods of about 12 years; assuming the same length of depletion period, the main difference between a medium and small reservoir appears to lie in the total amount of recoverable reserves and rate of output. (These estimates were, of course, based on economic and technological conditions existing at the time the report was written; they also do not include the effects of tertiary recovery methods.)

The recoverable portion of the total volume of oil in a reservoir depends on a number of factors. A primary determinant is the source of the reservoir's natural drive. Recovery factors differ from field to field,[12] with variations arising from the difference in the reservoir environment in which the hydrocarbons occur. In addition, the rate of extraction can affect the total volume ultimately

TABLE 2.1

BASIC DATA AND PRODUCING CHARACTERISTICS FOR A PETROLEUM RESERVOIR DRAINED BY A SINGLE 40-WELL PRODUCTION INSTALLATION

Basic Data	Large Reservoir	Medium Reservoir	Small Reservoir
Reservoir depth (feet)	8,000–10,000	8,000–10,000	8,000–10,000
Crude gravity	30° API	30° API	30° API
Barrels of oil recoverable per sq. km of reservoir volume	100,000	75,000	60,000
Area of reservoir (sq. km)	716 605	716 605	494 210
Average thickness of oil-bearing section within reservoir (feet)	150	75	50
Producing Characteristics			
Recoverable reserves (MMB)	175	65	25
Initial peak capacity (MB/D)	50	30	15
Years at peak capacity	3	2	1
Decline rate (per cent per year)	13	21	23
Depletion period (years)	20	12 [*sic*]	12

Note: The above conditions are intended to describe hypothetical, but nonetheless reasonable, situations; understandably, other situations or combinations of the variables are possible. In particular, the initial peak capacity numbers selected are near the maximum that could be expected for the reservoirs chosen. Many large, medium and small reservoirs cannot reach and sustain the producing rates assumed above.

Source: Adapted from National Petroleum Council, *Ocean Petroleum Resources* (Washington, D.C., 1975), Table 9, p. 32.

recovered, or 'ultimate recovery'. A concept often referred to in the petroleum industry is 'the maximum efficient rate of production' or M.E.R.; this is defined as the rate which, if exceeded, would lead to avoidable reduction in the volume ultimately recovered.

Thus, within the production life of a reservoir, if selling prices remain constant, a point will be reached where marginal costs may exceed marginal revenue[13] and production of that particular reservoir may no longer be profitable, i.e., uneconomic.

A common assumption is that the net price[14] of an exhaustible resource will rise along with the rate of interest.[15] This assumption will not hold if technological progress in extraction outpaces the rise in the rate of interest, if the nature of the discoveries results in lower costs of production and if substitutes are in sight over the long run.[16] It is therefore possible for real costs (and therefore for prices) to remain constant or even to decline over time, as will be shown in Chapter III.

OIL AND NATURAL GAS RESOURCES

Table 2.2 shows the estimated oil and gas reserves and potential resources in each of the South-East Asian countries. It will be noted that the largest estimates of potentially recoverable natural gas resources in the region are in Indonesia, while at least five countries are expected to have oil resources in the range of 10 to 100 billion barrels. In terms of proved reserves, Indonesia holds the largest for both oil and gas, while Brunei and Malaysia have gas reserves about one-third and two-thirds, respectively, of those found in Indonesia, and oil reserves of much smaller proportions.

Two significant developments have, however, occurred in two other countries in 1976, 1977, and 1978. Discoveries of gas in Thai territory in the Gulf of Thailand have resulted in estimates for proved reserves of 4.5 trillion cubic feet, while discoveries off Palawan in the Philippines have been estimated in some quarters to contain proved reserves of about 100 million barrels.[17] Although these discoveries are meagre in relation to the energy needs of these countries, which are at this writing about 95 per cent dependent on petroleum, these discoveries usher in a new era for them. The discoveries also follow years of exploration and numerous dry-hole experiences, and have provided encouragement for further exploration in the region, especially in the Philippines.

Discoveries off Vietnam in 1974 and 1975 also upgraded the petroleum potential of that country. Disruption of exploration work resulting from political developments unfortunately slowed down development of that potential. Reports of discoveries[18] in 1977 and 1978 in onshore areas near Hanoi (the Hanoi trough) and in the Mekong delta have been skimpy, but they nevertheless increase optimism on the future potential of Vietnam. Industry observers also consider the estimates in Table 2.2 on Vietnam extremely conservative in the light of these discoveries.

The resource estimates in Table 2.2 are sometimes regarded by industry observers to be generally high. Another approach—that tends to be controversial because it breaks away from 'conventional wisdom'—similarly gives an optimistic assessment of the petroleum potential of developing countries in general and of South-East Asia in particular. Grossling argues that the amount of geological/geophysical work and drilling in developing regions does not provide a sufficient basis for properly assessing the potential of their sedimentary areas.[19] He suggests that a large proportion ('roughly 1/2') of undiscovered world petroleum resources may be found in developing countries, about 12 per cent of which are in South Asia and South-East Asia.[20]

Table 2.3 shows outputs of oil and gas in the producing countries of South-East Asia for 1976 and 1977, as well as the projected outputs for certain countries. Indonesia is currently the largest oil producer in the region, Malaysia is expected to be a significant oil producer by 1990, and the Philippines is expected to be in the producer column well before then, with a projected production of about 145,000 barrels per day by 1990.

Brunei is currently the largest producer of natural gas, with Indonesia producing only slightly less. Natural gas, unless exploited locally, is marketable only by export in the form of liquefied natural gas (LNG). Brunei, Malaysia, and Indonesia have all proceeded to make arrangements to implement this technique; at present Brunei's output is piped to Sarawak (Malaysia); output from Malaysia and Brunei is then exported to Japan in the form of LNG.

Thailand, however, will consume output from its discovered fields domestically. Plans are under way to pipe the gas to the mainland to supply power generation plants and produce the accompanying by-products.

TABLE 2.2
ENERGY RESERVES AND RESOURCES, SOUTH-EAST ASIA

Country	Coal (10^6 m.t.)[a]		Commercial Fossil Fuels Petroleum (10^6 bbl.)		Natural gas (10^9 cu. ft.)		Hydro Power (assessed in MTCE)[d]	Uranium Recoverable (MTCE)[e]	Shale (10^6 m.t.)	Geothermal Power (10^6 watts)
	Reserves[b]	Resources[c] in place	Reserves[b] (1.1.79)	Potential Resources[c] Recoverable	Reserves[b] (1.1.79)	Potential Resources[c] Recoverable				
Brunei	—	—	1,480	D	8,000	D	—	—	—	—
Burma	13	286	45	C	150	C	1 384	—	U[f]	—
Cambodia (Kampuchea)	—	—	—	C	—	D	175	—	—	—
Indonesia	2 213	10 000[g]	10,200	C	24,000	B	923	—	—	8,000[h]
Laos	—	—	—	D	—	D	277	—	—	—
Malaysia	57[i]	400–500[j]	2,800	C	17,000	C	140	U[k]	—	—
Philippines	91	500[k]	100	C	—	C	121	U[k]	—	768[k] (1987)
Singapore	—	—	—	E	—	E	—	—	—	—
Thailand	235	NA	0.2	D	5,000	D	412	—	2 515[l]	U[m]
Vietnam	200[n]	1 000[n]	—	D[o]	—	D	1 077	—	—	—

Sources: World Energy Conference, Survey of Energy Resources (London: 1974), for coal and most sources other than oil and gas. Oil and Gas Journal, 25 December 1978, for oil and gas reserves. John Albers et al., Summary Petroleum and Selected Mineral Statistics for 120 Countries, Including Offshore, U.S. Geological Survey Professional Paper 817 (Washington, D.C.: U.S. Government Printing Office, 1973). Sherwood Frezon, Summary of 1972 Oil and Gas Statistics for Onshore and Offshore Areas of 151 Countries, U.S. Geological Survey Paper 885 (Washington, D.C.: U.S. Government Printing Office, 1974).

TABLE 2.2 (continued)

UN, ESCAP, 'Energy Resources in the Region', Document NR/WGMEPP/1, 6 July 1978.

UN, ESCAP, *Proceedings of the Twelfth Session of the Sub-Committee on Energy Resources and Electric Power*, Energy Resources Development Series, No. 11 (New York, 1974).

Indonesian Government National Committee, WEC, 1978 Energy Workshop.

Malaysian Government Delegation, UN, ESCAP Working Group Meeting on Energy Planning and Programming, Document NR/WGMEPP/CRP.2, 15 August 1978.

Philippine Government, Ministry of Energy, *Ten-Year Energy Development Plan, 1978–1987* (Manila: 1978).

Thailand Government, *Progress in Energy Development*, 1975.

[a] The figures cover black coal, brown coal and lignite (but not peat) with all the quantities from the tables in the WEC reference converted to standard black coal equivalent (7,000 kcal/kg) by the United Nations.

[b] *Reserves* refer to that subset of *resources* (see note c below) that not only have been identified by geological or engineering methods but can also be produced at a profit and can be legally extracted at the time of reporting.

[c] The term *resources* refers to concentrations of a mineral discovered, undiscovered, and surmised to exist in such form that extraction is currently or potentially feasible.

[d] UN, ESCAP (1978). The basis for the hydropower figures is the annual energy available at average river flow (load factor ranging from 30 to 80 per cent) taken from the WEC data, converted by the United Nations by a factor of 0.123 ton of equivalent coal per 1,000 kilowatt-hours, and multiplied by a commonly used figure of 50 years' availability. The figures are also supplemented by UN-collected data.

[e] UN, ESCAP (1978), and Philippine Government (1978). The economic limit of production cost at US$66 per kilogram was used by the UN. The Philippine write-up did not include any cost limit.

[f] WEC reports 217 megatonnes for Thailand and Burma, but see reference in i below.

[g] Indonesian Government. See *AWSJ*, 26 May 1978, and 1978 Workshop, National Committee, WEC.

[h] Indonesian National Committee, WEC 1978 Energy Workshop, speech by Mr. Subroto, 25 May 1978.

[i] UN, ESCAP (1974), p. 82.

[j] Malaysian Government (1978).

[k] Philippine Government (1978).

[l] UN, ESCAP (1974), p. 187.

[m] Thai Government (1975).

[n] UN, ESCAP (1974), p. 82.

[o] Industry observers consider this estimate too low, in view of discoveries made subsequent to estimate date. No new estimates are available at this writing.

Legend: for Petroleum and Natural Gas Resources
(10^9 bbl. of oil or 10^{12} cu. ft. of gas)

A = 1,000–10,000
B = 100– 1,000
C = 10– 100
D = 1– 10
E = 0.1– 1

— = Zero or negligible.
U = Unclear, unestimated, or unknown.
MTCE = Metric tons of coal equivalent.

Table 2.3 shows South-East Asia's increasing share, albeit small, of total world output of both crude oil and natural gas in the last decade. This appears to support the view that the potential of this region is only recently being tapped. Industry observers note that current production is coming from relatively mature fields—that is, production from Indonesia and Brunei—and that the undiscovered hydrocarbon potential of this region is significant within the context of the region.[21]

OTHER CONVENTIONAL ENERGY RESOURCES

Table 2.2 also shows the estimated reserves and potential resources for coal,[22] hydropower and uranium. Indonesia has the largest estimated recoverable coal resources in the region, and those of the rest appear to pale beside those of this country. Burma, Vietnam, and Indonesia have among the best hydropower resources in the world.[23]

Coal production in the region has increased at about 4 per cent per annum between 1972 and 1977, with most of the major increases taking place in the Philippines, Thailand, and Vietnam.[24] These outputs have also been supplemented by imports from sources such as China and Australia. In the meantime, however, Indonesia, the Philippines, and Thailand have taken steps towards developing new fields. These expansions have been favoured by the improved price relationships between coal and oil, and these expansions are expected to continue. The reasons for such expansions are not necessarily identical among the three countries. While the Philippines and Thailand seek to utilize their indigenous coal to reduce oil imports, Indonesia aims at substituting coal for oil to maintain its oil exports for the foreign exchange earnings required for the country's economic development programmes.

The ratio of utilized hydropower resources relative to potential is much higher in the region than in industrialized countries. With the current levels of oil prices, many South-East Asian countries, such as Indonesia, the Philippines, and Thailand, have accelerated the development of these water power resources; they also have plans to increase such utilization.[25]

The uranium resources of the countries of the region have not to

TABLE 2.3

PETROLEUM PRODUCTION: SOUTH-EAST ASIAN COUNTRIES, ACTUAL AND PROJECTED

	Crude Oil								Natural Gas					
	Actual Volume (10^3 b/d)				Compound Annual Change (%)			Projected Volume (10^3 b/d)	Actual Volume (10^9 cu. ft.)			Compound Annual Change (%)		Projected Volume (10^9 cu. ft.)
Country	1966	1971	1976	1977	1971/66	1976/71	1977/76	1990	1971[a]	1976[a]	1977[b]	1976/71	1977/76	1990
Brunei	95.0	128.1	203.9	207.0	+ 6.1	+ 9.7	+ 1.5	NA	44.0	298.8	306.6	+46.5	+ 2.6	NA
Burma	11.4	17.5	22.4	23.0	+ 9.0	+ 5.0	+ 2.7	NA	2.3	5.4	NA	+18.6	NA	NA
Indonesia	464.2	882.1	1,507.7	1,690.0	+13.7	+11.3	+12.1	2,500[c]	44.5	126.4	139.6	+23.5	+10.4	≥350[e]
Malaysia	1.0	65.5	165.9	190.0	+130.0	+20.0	+14.5	950[d]	2.3	22.6[c]	NA	+57.5	NA	500[e]
Philippines	—	—	—	—	—	—	—	145[f]	—	—	—	—	—	—
Thailand	.04	0.3	0.2	0.2	+ 50.0	−33.0	0	0.8	—	—	—	—	—	165[g]
Total S.E.A.	571.6	1,093.5	1,900.1	2,110.2	+13.9	+11.7	+11.1	NA	53.1	453.0	NA	—	—	—
World	33,869.2	49,589.8	58,059.0	59,527.8	+ 7.9	+ 3.2	+ 2.5	NA	40,810.7	49,300.7	NA	+ 4.6	NA	NA
S.E.A. as % of World	1.7	2.2	3.3	3.5	—	—	—	—	0.1	0.9	NA	—	—	—

Source: Oil and Gas Journal, 26 December 1977 issue for 1977 oil data. U.S., Bureau of Mines, annuals, for 1976 gas and oil data. UN, ESCAP, for 1977 gas output.

[a]U.S. Bureau of Mines marketed production data used.

[b]Data recomputed from UN, ESCAP figures except for Malaysia. This was not entered because the series was not comparable with the B.M. series.

[c]Estimated. See *Asian Wall Street Journal*, 23 March 1977, p. 10; *AWSJ*, 20 September 1978, pp. 1, 12; UN, ESCAP Document NR/WGMEEPP/CRP.2, 15 August 1978.

[d]Using 1980 optimistic projection of Scottish Council and advancing it to 1990.

[e]Data on Sarawak only. No data on joint output from oil wells in Sabah were available.

[f]Philippines, Ministry of Energy.

[g]Thailand, Natural Gas Organization of Thailand. Based on design capacity of 500 MMCFD by 1981 and an assumed output of 450 MMCFD.

NA = Not available

date been completely identified. The Philippines has made some modest discoveries and both the Philippines and Thailand are continuing their search for commercial-sized local deposits.[26] Nuclear power plants have, however, been scheduled for construction in the Philippines and Indonesia.[27] A modest uranium recovery was expected in central Philippines in a mine which was scheduled to be producing by 1979.[28]

NEW ENERGY SOURCES

Countries in the region have begun to explore the use of non-traditional sources such as oil shale and geothermal resources. Thailand has for some time been studying the possibilities of using its vast oil shale reserves. To date the costs of developing these reserves have prevented their utilization.[29]

The Philippines has, however, successfully implemented the development of its geothermal resources.[30] In 1978 a geothermal plant in central Philippines was already operating, while construction of others was in progress. When fully developed, its 25 potential areas are expected to have a total capacity of 768 megawatts.[31]

NON-COMMERCIAL SOURCES

Except for Singapore, which is essentially urban, most countries of the region have what are termed 'non-commercial'[32] forms of energy such as fuelwood, agricultural wastes (e.g., bagasse from sugar mills) and animal wastes (e.g., cow-dung). The large-scale use of fuelwood is increasingly being condemned, however, because of the environmental impacts of deforestation.[33]

The solar energy potential of the region is high, and countries such as Singapore and the Philippines are exploring ways of harnessing this energy for industrial and commercial use. Tropical areas have, in general, comparatively low average wind velocity that could be utilized on a continuous basis, although there are areas like Thailand where simple handmade windmills are used for low head-water pumping.[34]

Demand for Petroleum and Other Energy Resources

In general, a primary energy resource is converted into a secondary form to increase the versatility of its end-uses. Certain end-uses result in more efficient thermodynamic utilization of primary forms, e.g., natural gas for cooking as opposed to its use in electric power generation. In the final analysis, the demand for a specific energy resource is determined by the cost of using such fuel form. Basically, such demand depends on the cost of alternative fuels, the availability of such fuels, and developments in technology that influence such use.

HISTORICAL CONSUMPTION PATTERNS

Table 2.4 shows the patterns of consumption for the period 1966 to 1976 in South-East Asia of the four conventional sources of energy that are sometimes classed as 'commercial'. In addition, separate information concerning consumption of fuelwood is included.

The data show that in most countries oil has increasingly provided the major portion of the energy requirements of South-East Asia. The exceptions are Brunei, Burma, and Vietnam. Brunei has increasingly relied on its natural gas output, while Burma and Vietnam have increased their dependence on coal.

The analysis looks at two five-year periods. The year 1971 is taken as the intermediate year, with 1966 consumption reflecting usage pattern before the 1973 escalation in oil prices. While oil and gas consumption in Brunei, Indonesia and Malaysia—the major producers in the region—increased in the second period, a perceptible decline in the growth rate of petroleum occurred in the oil-importing countries of the region during the period 1971–6. At the same time, except for the producing countries, growth of total commercial energy consumed also slowed down during the second period.

For the first time since the United Nations started its Statistical Series J, issue number 20 (covering the period 1971–5 for consumption data) contained information on fuelwood consumption. As mentioned earlier, coal and its various forms, liquid hydrocarbons (such as crude petroleum and natural gas liquids), natural gas, and electricity are traditionally referred to as commercial energy, and comprehensive statistics on production and consumption of these energy forms are available. Fuelwood, agricultural

TABLE 2.4

PATTERNS OF CONSUMPTION OF PRIMARY ENERGY IN SOUTH-EAST ASIA, 1966–1976 (SELECTED YEARS)

(Volume in thousand metric tons of coal equivalent)

Country and Energy Form	1966 Volume	1966 Per Cent of Total	1971 Volume	1971 Per Cent of Total	1976[c] Volume	1976[c] Per Cent of Total	Compounded Annual Growth Rate (per cent) 1971/66	Compounded Annual Growth Rate (per cent) 1976/71
BRUNEI								
Coal and lignite[a]	—	—	—	—	—	—	—	—
Crude oil and NG liquids	53	22	101	29	120	5	13.8	2.3
Natural gas	188	78	245	71	2 400	95	5.4	58.0
Hydro-nuclear electricity	—	—	—	—	—	—	—	—
Total commercial energy	241	100	346	100	2 520	100	7.5	48.5
Fuelwood[b]	NA	—	16	—	20 (1975)	—	NA	4.8
% of commercial energy	—	NA	—	5	—	1	—	—
BURMA								
Coal and lignite[a]	95	7	227	13	220	14	19.0	-5.7
Crude oil and NG liquids	1 190	89	1 466	84	1 240	82	4.2	-3.3
Natural gas	15	1	9	—	NA	—	—	NA
Hydro-nuclear electricity	34	3	50	3	60	4	-10.0	NA
Total commercial energy	1 334	100	1 752	100	1 510	100	5.6	-3.0
Fuelwood[b]	NA	—	4 520	—	4 750 (1975)	—	NA	1.3
% of commercial energy	—	NA	—	258	—	315	—	—

Country and Energy Form	1966 Volume	1966 Per Cent of Total	1971 Volume	1971 Per Cent of Total	1976[c] Volume	1976[c] Per Cent of Total	Compounded Annual Growth Rate (per cent) 1971/66	1976/71
CAMBODIA (KAMPUCHEA)								
Coal and lignite[a]	37	11	10	6	NA	—	-23.0	NA
Crude oil and NG liquids	302	89	162	92	130	100	-12.0	-4.5
Natural gas	—	—	—	—	—	—	—	—
Hydro-nuclear electricity	—	—	3	2	—	—	—	—
Total commercial energy	339	100	175	100	130	100	-12.5	-5.5
Fuelwood[b]	NA	—	915	—	1 010 (1975)	—	NA	2.5
% of commercial energy	—	NA	—	523	—	777	—	—
INDONESIA								
Coal and lignite[a]	339	3	230	1	210	1	-7.5	-2.0
Crude oil and NG liquids	6 491	58	11 462	70	25 640	84	12.0	17.5
Natural gas	4 212	38	4 571	28	4 360	14	1.7	-1.0
Hydro-nuclear electricity	100	1	175	1	220	1	11.8	4.7
Total commercial energy	11 142	100	16 438	100	30 430	100	8.1	22.0
Fuelwood[b]	NA	—	25 000	—	27 750 (1975)	—	NA	2.7
% of commercial energy	—	NA	—	152	—	91	—	—

TABLE 2.4 (continued)

Country and Energy Form	1966 Volume	1966 Per Cent of Total	1971 Volume	1971 Per Cent of Total	1976c Volume	1976c Per Cent of Total	Compounded Annual Growth Rate (per cent) 1971/66	1976/71
LAOS								
Coal and lignite^a	—	—	—	—	—	—	—	—
Crude oil and NG liquids	120	100	209	97	200	95	11.7	-1.0
Natural gas	—	—	—	—	—	5	NA	7.4
Hydro-nuclear power	—	—	7	3	10	—	NA	-0.5
Total commercial energy	120	100	216	100	210	100	12.5	
Fuelwood^b	NA	—	690	—	740 (1975)	—	NA	1.8
% of commercial energy	—	NA	—	319	—	352	—	—
MALAYSIA								
Coal and lignite^a	23	1	58	1	50	1	20.5	-3.0
Crude oil and NG liquids	3 753	95	4 578	95	7 130	93	4.1	9.2
Natural gas	80	2	67	1	320	4	-3.5	35.0
Hydro-nuclear electricity	85	2	129	3	120	2	8.7	-1.5
Total commercial energy	3 941	100	4 832	100	7 620	100	4.2	9.5
Fuelwood^b	NA	—	1 282	—	1 370 (1975)	—	NA	1.7
% of commercial energy	—	NA	—	27	—	18	—	—

Country and Energy Form	1966 Volume	1966 Per Cent of Total	1971 Volume	1971 Per Cent of Total	1976c Volume	1976c Per Cent of Total	Compounded Annual Growth Rate (per cent) 1971/66	Compounded Annual Growth Rate (per cent) 1976/71
PHILIPPINES								
Coal and lignite[a]	104	1	54	—	180	1	−12.0	3.2
Crude oil and NG liquids	7 416	96	11 284	97	13 610	95	8.7	3.8
Natural gas	—	—	—	—	—	—	—	—
Hydro-nuclear electricity	187	3	317	3	600	4	11.2	13.6
Total commercial energy	7 707	100	11 655	100	14 390	100	8.6	4.3
Fuelwood[b]	NA	—	4.850	—	5 550 (1975)	—	NA	3.4
% of commercial energy		NA	—	42	—	39	—	—
SINGAPORE								
Coal and lignite[a]	6	—	5	—	NA	NA	−3.5	NA
Crude oil and NG liquids	1 774	100	3 516	100	5 150	100	14.7	7.9
Natural gas	—	—	—	—	—	—	—	—
Hydro-nuclear electricity	—	—	—	—	—	—	—	—
Total commercial energy	1 780	100	3 521	100	NA	100	14.6	NA
Fuelwood[b]	NA	—	11	—	10 (1975)	—	NA	−2.5
% of commercial energy		NA	—	—	—	—	—	—

TABLE 2.4 (continued)

Country and Energy Form	1966 Volume	1966 Per Cent of Total	1971 Volume	1971 Per Cent of Total	1976c Volume	1976c Per Cent of Total	Compounded Annual Growth Rate (per cent) 1971/66	Compounded Annual Growth Rate (per cent) 1976/71
THAILAND								
Coal and lignite[a]	61	2	162	1	260	2	21.5	9.9
Crude oil and NG liquids	3 848	95	11 131	97	12 490	95	23.5	2.3
Natural gas	—		—		—		—	—
Hydro-nuclear electricity	131	3	246	2	470	3	13.4	13.8
Total commercial energy	4 040	100	11 539	100	13 220	100	23.5	2.2
Fuelwood[b]	NA	—	3 688	—	3 950 (1975)	—	NA	1.7
% of commercial energy	—	NA	—	32	—	30	—	—
VIETNAM								
Coal and lignite[a]	3 042	36	2 551	23	4 800	83	-3.5	13.5
Crude oil and NG liquids	5 431	64	8 671	77	910	16	9.8	-45.0
Natural gas	—		—		—		—	—
Hydro-nuclear power	49	—	76	—	70	1	9.2	-1.5
Total commercial energy	8 522	100	11 298	100	5 780	100	—	—
Fuelwood[b]	NA	—	3 988	—	4 080 (1975)	—	NA	0.5
% of commercial energy	—	NA	—	35	—	71	—	—

[a] Includes peat.
[b] Conversion ratio used: 4 cu. m greenwood equivalent to 1 ton coal (cf. ESCAP Document, p. 45).
[c] Rounded off to ten thousandth unit.

Source: United Nations, *World Energy Supplies*, Series J. No. 19 and No. 20, and ESCAP Document NR/WGMEPP/1, 6 July 1978.

Legend: — = Zero or negligible.
NA = Not available.
NG = Nat...

and animal wastes, solar energy, and wind energy are, on the other hand, referred to as non-commercial energy forms. At this time, statistics on production and consumption are available only for fuelwood. When combined with the usage of animal dung in some countries, usage of non-commercial forms of energy is estimated to account for a significant portion of energy consumption (probably as much as 50 per cent of total in some cases).[35]

In Table 2.4 data on fuelwood consumption in South-East Asian countries are included to provide a better perspective of the energy consumption patterns in these countries than had heretofore been possible. It will be noted that in the relatively less industrialized countries in the region—Burma, Cambodia, Indonesia, and Laos—fuelwood has provided more energy than the traditional, commercial forms. There are suggestions that the increasing costs of petroleum products may increase the use of non-commercial forms in the developing countries of the Far East.[36] Such increase may not be significant, however, in view of their limited application in the major economic activities related to large-scale economic development programmes.

FUTURE DEMAND FOR PETROLEUM IN SOUTH-EAST ASIA

The demand for energy is a derived demand. In South-East Asian countries as elsewhere, this demand primarily results from the demand for public as well as private goods. Because energy consumption is related to increased productivity, either by supplementing or complementing human and animal energy, energy consumption is very closely related to economic growth. Energy consumption will tend to increase faster in developing rather than in developed countries, because the potential for replacing human and animal energy with electrical or mechanical forms is larger.[37] South-East Asia is considered to be one of the fastest growing regions of the world today, and for that reason the demand for energy in its commercial forms may be expected to rise at rates above those expected for both the developed or developing areas as a whole.

Although there is a definite trend towards greater diversification of energy sources in South-East Asia, dependence on petroleum for most of the requirements of the region may be expected to be the norm in the medium term. Both economic and technical factors dictate this. First, there are long lead times for developing alternative sources, even when present. Second, energy substitution may not be technically possible with industrial or agricultural equipment available or in use. Third, cost/price relationships still favour the use of petroleum, even when imported at present prices.

Research is progressing on the utilization of agricultural wastes and other forms of biomass as energy resources. Although the potentials of adopting this new source of energy appear to be significant, its impact on South-East Asia's energy market may not be dramatic in the next ten or fifteen years.

Table 2.5 shows projections of energy and petroleum requirements for the years 1985 to 1990 in three countries—Indonesia, the Philippines, and Thailand.[38] These countries have undertaken serious energy development programmes for various reasons. Both the Philippines and Thailand are heavily dependent on imports, and their oil import bills have since 1974 represented from 25 to 30 per cent of their export earnings.[39] Their recent discoveries will only partly alleviate the high foreign exchange costs of their energy requirements, and they have embarked on programmes to increase their use of alternative sources, such as hydro and nuclear power, geothermal energy, coal, oil shale, etc. Indonesia, the largest oil producer in the region and a member of the Organization of Petroleum Exporting Countries (OPEC), is concerned about its energy usage because it wants to assign to its oil and gas resources a major role as a foreign-exchange earner. These foreign-exchange reserves are necessary to finance its massive economic development programme in the next decade or so. Indonesia, therefore, seeks to develop its other energy resources such as its hydropower and coal resources to reduce the domestic demand for its petroleum resources.

Notwithstanding such efforts at diversification, as Table 2.5 shows, estimates based on projections by official groups in these three countries project the share of oil and gas to be anywhere from 60 to about 80 per cent of the countries' total energy requirements in 1990. No projections are available for the other countries, but it

TABLE 2.5

PROJECTIONS OF ENERGY AND PETROLEUM DEMAND, 1985 AND 1990: INDONESIA, THE PHILIPPINES AND THAILAND

Country	Energy Volume		Oil				Natural Gas			
	1985	1990	1985		1990		1985		1990	
	10⁶ MTCE	10⁶ MTCE	10⁶ MTCE	% of Energy	10⁶ MTCE	% of Energy	10⁶ MTCE	% of Energy	10⁶ MTCE	% of Energy
Indonesia	52.6[a]	82.4[a]	43.0[b]	81.8	64.3[c]	78.0	5.4[b]	10.3	7.4[c]	9.0
Philippines	34.4[d]	51.5[d]	25.4[e]	73.2	30.9[f]	59.6	—	—	—	—
Thailand	31.2[g]	50.2[h]	19.7[g]	63.0	34.1[g]	68.0	4.9[g]	15.7	6.5[g]	13.0[g]

Sources: Estimated from analyses and data in: Papers from Workshop organized by the Indonesian National Committee of the World Energy Conference, 25–26 May 1978, Jakarta;

Indonesian National Committee, World Energy Conference, *Hasil-hasil Lokakarya Energi: Perkiraan Kebutuhan Energi Indonesia, 1975–2000* (Jakarta: 12–13 May 1977);

Budi Sudarsono, 'Indonesian Energy Data and Projections' in *Atom Indonesia*, Vol. 2, No. 1 (January 1976);

Budi Sudarsono et al., *Proyek Penelitian Perspektip Jangka Panjang Perekonomian Indonesia* (Jakarta: LEKNAS-LIPI, 1977).

Philippines, Ministry of Energy, *Ten-Year Development Program, 1978–1987* (Manila: 1977);

Thailand, UN Delegation, *Energy in Thailand* (Bangkok: 1978).

[a] Based on 'probable' forecast in *Hasil-hasil Lokakarya Energi*.

[b] Using proportions in Sudarsono et al., Table XI.

[c] Assuming increasing use of coal by 1990, a declining share for gas, and increased share of hydroelectric power and other sources.

[d] Estimated, using 8.4 per cent growth rate, interpolating and extrapolating projections by Ministry of Energy.

[e] Estimated, using 4.2 per cent growth rate, interpolating projections of Ministry of Energy.

[f] Estimated, using 4.0 per cent growth rate, assuming reduced growth rate for oil as geothermal and other resources are developed.

[g] Estimated, using projections and assumptions in Thailand Government report (1977), Table 1–3.

[h] Estimated by interpolation of estimates in Thai report.

Note: Conversion factors used:

1 kcal × 10⁶ = 0.14 MTCE

1 bbl. of oil = 0.21 MTCE (Ref: WAES Report, Table 1–1, p. 64)

MTCE = Metric ton of coal equivalent.

would be safe to assume that petroleum will continue to form a major source of energy to fuel their economic development programmes despite the development of other alternatives. (See, for example, the development of hydroelectrical energy resources in Peninsular Malaysia,[40] as well as plans by Malaysia to make use of natural gas jointly produced with oil in Sabah and Sarawak in power generation.[41] See also Singapore's plans to switch to coal fuel in its new power station that will have a designed capacity of 2,100 megawatts.[42])

These projections may not, of course, be viewed independently of economic and energy developments within and outside the region. The projections themselves are made on the assumptions of certain economic growth rates, which are dependent on growth in the developed countries—the markets for the exports of South-East Asian countries. GNP growth rates in South-East Asia were above 6 per cent in 1975–6 and at an average of 5 to 10 per cent during the period 1967–76.[43] It may be reasonable to assume that average growth rates at 6 per cent will continue up to 1990, while the economies of the developed countries grow at about 4 per cent.[44] While energy consumption in developing countries as a whole may be assumed to grow at about 6 per cent,[45] the higher projections in the relatively more industrialized and faster-growing countries of South-East Asia (which assume income elasticities at around 1.5) do not appear out of line.[46]

Varying forecasts have been made of the global demand-supply situation by 1985 and 1990.[47] Nevertheless, there appears to be no disagreement on the continued importance of petroleum as a source of energy. Recent forecasts of the shares of oil and gas by Exxon Corporation and the World Bank indicate close to 50 per cent dependence on oil and about 15 per cent dependence on natural gas globally by 1990.[48]

	World Bank	Exxon
Oil	46%	48%
Gas	16%	15%

The Exxon estimate was based on a growth rate of only 2.5 per cent annually for oil during the period 1980 to 1990, while non-oil sources grow at 5.2 per cent.

Given these premises then, it may be expected that the South-East Asian countries will want to see their petroleum resources explored and discovered. Even for a country like Malaysia, which has expressed that it is in no great hurry to see its petroleum resources developed—since it has other products like tin, rubber, and palm oil that bring in foreign exchange—there is some concern about the life of its present reserves.[49] Thus it is reasonable to state that nations with potentially recoverable petroleum have a demand function for petroleum reserves that can be satisfied at some point by a supply function for such reserves.

1. The term 'demand' will be used synonymously with 'consumption' in this study.

2. *Measured* resources are those whose quality and quantity have been estimated, within a margin of error of less than 20 per cent, from analyses and measurements done for closely spaced and geologically well-known sample sites. The term *indicated* is applied to resources whose quantity and quality have been estimated partly from sample analyses and partly from reasonable geologic projections. The term *inferred* refers to materials in unexplored but identified deposits estimated on the basis of geologic evidence.

3. The term *hypothetical* resources is used to refer to undiscovered resources that may be expected to exist in a discovered area under known geologic conditions, while the term *speculative* resources refers to resources that may occur in areas where no discoveries have yet been made, both in known and unknown types of deposits.

4. These definitions are taken from Federal Power Commission, *National Gas Survey* (Washington, D.C.: 1974), Appendix A, Vol. I, Chapter 9.

5. The notion of *recovery efficiency*, expressed in percentage terms, refers to the quotient of the total proved petroleum at a point in time divided by the estimated cumulative total in reservoirs known to exist at that point in time.

6. The relationships between resource base and extraction of reserves may be expressed in a more formal way as follows: There is an estimated quantity of oil-in-place, whose production is constrained by engineering and economic factors, and this may be represented as follows:

$$(2.1) \quad \int_0^T q(t)dt \leqslant xS$$

where S represents original oil-in-place, x is a factor indicating the maximum proportion of the oil-in-place that is physically recoverable with existing technology, T is the production time horizon, and $q(t)$ the rate of production. The unknown function $q(t)$ may be expressed as a function of the initial capacity installed and the production decline rate such that:

$$(2.2) \quad q(t) = q_0 e^{-at},$$

where o is the subscript indicating the present, t is any point in time, q_0 represents initial installed capacity, and a represents the rate of decline in production. As described above, total resource recovery is generally a negative function of the rate of production. It may thus be postulated that reserves, S_0, may be expressed as:

$$(2.3) \quad S_0 = xS - \beta q_0 e^{-a} - \gamma q_0,$$

where β and γ are physical parameters related to geological conditions, q_0 is the installed capacity, and $q_0 e^{-a}$ is the initial rate of production. The term 'installed capacity' refers to the rate of production that is sustainable for a brief period without further drilling of new wells.

See R. J. Kalter *et al.*, *Atlantic Outer Continental Shelf Energy Resources: An Economic Analysis* (Ithaca, New York: Cornell University), A.E. Res. 74–17, mimeographed, and W. F. Lovejoy and P. T. Homan, *Economic Aspects of Oil Conservation Regulation* (Baltimore: Johns Hopkins University Press for the Resources for the Future, 1967).

7. The term *secondary recovery* is used when oil is recovered by supplementing the natural reservoir pressure by other agents (such as water flooding or gas injection).

8. The term *tertiary recovery* refers to any process used to obtain recovery levels not achievable with conventional primary and secondary processes.

9. The term 'original oil-in-place' is used to refer to the volume of crude oil estimated to be in known reservoirs prior to any production.

10. See American Petroleum Institute, *Reserves of Crude Oil, Natural Gas Liquids, and Natural Gas in the United States and Canada* (Washington, D.C.: 1978).

11. National Petroleum Council, *Enhanced Oil Recovery* (Washington, D.C.: 1976), p. 12.

12. A *field* is the general area underlain by one or more reservoirs formed by a common structural or stratigraphic feature. R. E. Megill, *Exploration Economics* (Tulsa, Oklahoma: The Petroleum Publishing Company, 1971), p. 12.

13. The term *marginal cost* refers to incremental cost or the cost of producing an additional unit; the term *marginal revenue* refers to the revenue added by an additional unit produced.

14. The difference between production costs (which include a minimum, or reasonable, return on investment—or 'normal' profit) is referred to in economic jargon as *economic rent*. Other authors use the term 'net price'.

15. See H. Hotelling, 'The Economics of Exhaustible Resources', *Journal of Political Economy*, Vol. 39 (April 1931), pp. 137–75; R. L. Gordon, 'A Reinterpretation of the Pure Theory of Exhaustion', *Journal of Political Economy*, Vol. 75 (June 1967), pp. 274–86; O. C. Herfindahl, 'Some Fundamentals of Mineral Economics', *Land Economics*, Vol. 31 (May 1955), pp. 131–8; A. D. Scott, 'The Theory of the Mine Under Conditions of Certainty', in M. Gaffney (ed.), *Extractive Resources and Taxation* (Madison: University of Wisconsin, 1967), pp. 25–62; and R. Solow, 'Intergenerational Equity and Exhaustible Resources', *Review of Economic Studies* (December 1974), pp. 29–46.

16. See, for example, W. Nordhaus, 'The Allocation of Energy Resources', *Brookings Papers on Economic Activities*, No. 3 (Washington, D.C.: Brookings Institution, 1973). See also P. Dasgupta and J. E. Stiglitz, *Uncertainty and the Rate of Extraction Under Alternative Institutional Arrangements*, Technical Report No. 179, Institute for Mathematical Studies in the Social Sciences (Stanford: Stanford University, 1976), p. 3.

17. See the *Oil and Gas Journal*, 25 December 1978. Henceforth the symbol *OGJ* will be used to refer to this publication.

18. See *Petroleum Economist*, September, 1977, p. 373; *OGJ*, 31 July 1978, p. 102, 'Viet Nam claims more Hanoi trough discoveries', and *Petroleum News Southeast Asia*, September 1978, p. 3. (Henceforth, the symbols *PE* and *PN* will be used to refer to these publications.)

19. To gauge the potential of given areas, one needs to know the sedimentary area, sedimentary volume, thicknesses, lithology, structure, etc., but these data may only be obtained from geological and geophysical exploration and drilling. See B. F. Grossling, 'The Petroleum Exploration Challenge with Respect to the Developing Nations', in R. Meyer (ed.), *The Future Supply of Nature-Made Petroleum and Gas*, Proceedings UNITAR-IIASA Conference, Laxenburg, Austria, 5–16 July 1976 (Pergamon Press, 1977), pp. 57–69.

20. Grossling estimates that in the developing regions, about 34 per cent of undiscovered petroleum is in Latin America, 35 per cent in Africa and Madagascar, about 25 per cent in South and South-East Asia and about 6 per cent in mainland China. (See page 8 and Table 8 in B. F. Grossling, 'A Long-Range Outlook of World Petroleum Prospects', prepared for the Subcommittee on Energy of the Joint Economic Committee, Congress of the United States, 2 March 1978.) He defines South and South-East Asia as the mainland countries east of the Middle East and south of China plus the islands of Oceania excluding Japan, Australia and New Zealand. See Grossling (1977), op. cit., p. 59.

21. See, for example, comments at the Circum-Pacific Conference, Honolulu, U.S.A., July–August 1978, reported in *OGJ*, 28 August 1978, pp. 37–8, 'Esso: Far East oil potential limited'.

22. The term 'coal' is used here to cover all kinds of coal and lignite but not including peat.

23. See World Energy Conference, *World Energy Survey* (1974).

24. See United Nations, ESCAP, 'Energy Resources in the Region', Document NR/WGMEPP/1, 6 July 1978, Annex II.

25. See papers presented at the Indonesian National Committee, World Energy Conference, *1978 Workshop on Energy*, Jakarta, 26–27 May 1978; the Philippines' Ministry of Energy, *Ten-year Energy Development Program, 1978–1987* (Manila, 1978); and the Thailand Delegation to the UN, ESCAP Committee on Natural Resources, *Progress in Energy Development in Thailand* (Bangkok: October 1975).

26. See the Philippines' Ministry of Energy report (1978), op. cit., p. 80, and UN, ESCAP (1978) document.

27. See *Asian Wall Street Journal* (Hong Kong), 19 December 1978, p. 3, and Philippines, Ministry of Energy (1978), op. cit. (Henceforth, the *Asian Wall Street Journal* will be cited as *AWSJ*.)

28. See the Ministry of Energy (1978), op. cit.

29. Thailand Delegation to UN, ESCAP (1975), op. cit.

30. Its policies are spelled out in Philippines Act 1442 on geothermal exploration.

31. Ministry of Energy (1978), op. cit., p. 78.

32. The term 'commercial' is used in the energy literature generally to refer to forms traded in large markets, and the term 'non-commercial' to those not falling under the former category.

33. See, for example, the unpublished papers read at the 1978 Workshop on Energy in Jakarta.

34. UN, ESCAP (1978), p. 11.

35. See UN, ESCAP (1978), p. 3.

36. See UN, ESCAP (1978), p. 3.

37. The historical energy/GDP relationships in South-East Asian countries is discussed in C. M. Siddayao, *The Off-shore Petroleum Resources of South-East Asia* (Kuala Lumpur: Oxford University Press, 1978), Chapter I.

38. These projections are based on the underlying assumption of a relationship between economic growth (as represented by the Gross National Product), population growth, and demand for energy. Thus, the estimating equation in the Indonesian case was in the form, $\log E/P = a + b \log Y/P$, where E = total energy consumption, P = population, and Y = gross national product. See Budi Sudarsono *et al.*, *Proyek Penelitian Perspektip Jangka Penjang Perekonomian Indonesia* (Jakarta: LEKNAS-LIPI, 1977). In the Philippine case, the energy-GNP relationship was expressed as $\log C = \log A + \alpha \log \bar{Y} + \beta \log \bar{P}$. See Ministry of Energy, *Ten-Year Energy Development Program, 1978–1987* (Manila: 1978). The Thai case was not specified in the report, but implied in Table I–3, of the Thai Government, UN Delegation's report, *Energy in Thailand* (Bangkok: 1978). Explanations of the reasons for assigning specific percentages to individual energy forms are argued in the respective reports.

39. Central Bank of the Philippines, *Statistical Bulletins* and *Annual Reports*, and International Monetary Fund, *International Financial Statistics*.

40. *PN*, 'Hydropower in Peninsular Southeast Asia', June 1977, p. 16.

41. *PN*, July 1978, p. 18, and *AWSJ*, 15 June 1978, p. 2.

42. *Straits Times* (Singapore), 3 September 1977, p. 8, and 22 October 1977, p. 6. (Henceforth, this reference will be cited as *ST*.)

43. Asian Development Bank, *Key Indicators of Developing Member Countries of ADB*, April 1977, and the World Bank *Atlas*, 1976, as presented in Chia Siow Yue,

'Economic Developments in the Region', in *Southeast Asian Affairs 1978* (Singapore: Institute of Southeast Asian Studies, 1978).

44. See assumption by the World Bank, in *Price Prospects for Primary Commodities*. Report No. 814/78, June 1978, p. 277.

45. Ibid.

46. The elasticity concept is a device to indicate the degree of responsiveness of one variable to changes in another variable. The income elasticity of energy demand refers to the relationship between relative changes in energy consumption resulting from relative changes in real GNP. In symbolic form, $e_y = \%\Delta y / \%\Delta x$.

47. See, for example, the reports by the U.S. Central Intelligence Agency, 'The International Energy Situation: Outlook to 1985', April 1977; by the International Energy Agency, *Towards an International Strategy for Energy Research and Development*, September 1977, and the Workshop on Alternative Energy Strategies (WAES), *Energy: Global Prospects, 1985–2000*, 1977.

48. World Bank, Report 814/78, op. cit., Table 4, p. 290, and Exxon Corporation, 'World Energy Outlook', January 1977, Chart 5.

49. Conversation with Petronas official. This view has also appeared in the press on many occasions.

III

Petroleum Exploration: Some Distinguishing Characteristics of and Structural Changes in the Region

THE exploration and development picture in South-East Asia has undergone dramatic changes over the last ten years. These have been the result of structural changes both within and outside the region. In order to understand the importance of these changes, the basic variables influencing the supply function of petroleum reserves will first be briefly summarized. The structural changes that have influenced the level and kind of activity in South-East Asia will then be discussed. Finally, developments in the Indonesian legal framework—which, in most respects, have set the pattern for exploration contracts in all other South-East Asian nations—will be traced, as a preliminary to presenting a comparative outline of contractual arrangements in the region in Chapter IV.

The Supply Function for Petroleum Reserves[1]

FACTORS INFLUENCING EXPLORATION AND DEVELOPMENT

The petroleum industry is an extractive industry. The 'upstream' phase[2] involves three stages: *exploration*, *development*, and *production*. Figure 3.1 presents a diagrammatic scheme of these stages up to the time petroleum is transported for processing or marketing in its primary form. In a significant sense, the second phase of the exploration decision—exploratory drilling—largely determines the direction of the upstream process, although important decisions are made at each stage of the industry.

The decision to drill exploratory wells is the first step in a series

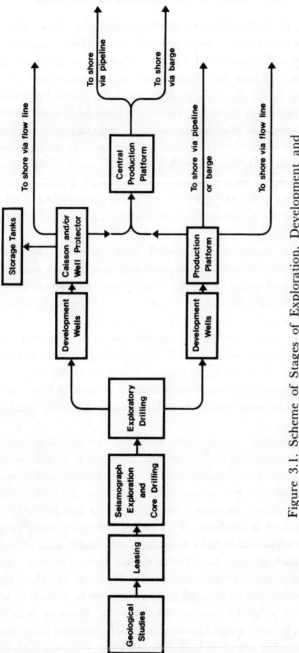

Figure 3.1. Scheme of Stages of Exploration, Development and Production of Petroleum Resources

of decisions to invest huge sums of money in a certain area. The mere physical presence of oil and gas resources does not guarantee that they will be produced. Certain technical and economic conditions must favour the exploration and development of those resources. Since decisions must be made to spend huge sums of money before all facts are known, a third and most important single variable in the assessment of an exploratory investment is risk.[3] The economic decision in exploration may be expressed in terms of the variables influencing that decision, i.e., the oil or gas prospects of an area as evaluated in geological and geophysical studies, the costs and gains from exploratory drilling and later developing the potential reservoir, and the probability of losing out on the venture.

Technical considerations include the potential presence of resources identified by geological and geophysical studies, the availability of technology and infrastructure to conduct the type of exploration required and the development of discovered resources, and the natural environment of the discovery.

Economic considerations involve the size of the expected output, price expectations over the production period, the costs of exploration, development and production, other costs that may be incurred in connection with operations, and various forms of payments to the government (royalties, taxes, bonuses, production shares, etc.).

Risk elements include both technical (geological/geophysical) and non-technical risks. Non-technical risks may include commercial (economic) and political risks.[4] Although the technology for detecting oil and gas accumulations has been improving, no one has found a better alternative to drilling into a trap to ascertain the existence and the volume of oil or gas resources within a given area. In areas which have been both intensively and extensively drilled (such as some parts of the United States), sizes of accumulations may be estimated statistically. The 'success ratios' (or the ratio of successful wells to total exploration wells drilled) in known areas may be used to project the chance of discovery in adjacent areas. In frontier areas, however, which characterize many parts of the developing world including South-East Asia, minimal knowledge drastically reduces the usefulness of statistical extrapolations. In addition to basic geological risks, recent history suggests there are risks related to the truncation of the operation following discovery and to recovering the expected return on one's investment. This leads us to the next subset of variables.

Legal and *administrative* considerations would normally be included in the economics variable, but in a country where it is difficult to predict changes in the institutional framework within which the producer must operate, the legal/administrative factor may be included in the risk element.

THE COST FUNCTION OF PETROLEUM RESERVES

The supply curve is generally discussed in terms of the relationship between the price of a commodity and the quantity supplied. This relationship holds if all other variables which determine the supply of any commodity are unchanged. The supply function of petroleum reserves in South-East Asia is complex. Nevertheless, the basic factors underlying that supply function remain the same. These are the items that determine the costs of producing the commodity, in this case, petroleum reserves, and the price the producer will receive, or his revenues. Expressed another way, underlying the supply function are the factors that determine the cost function and the revenue function. Assuming we know the elements in the cost function, which includes the rate of return to capital, we can express the supply function in terms of costs.

There are valid reasons for taking this approach. In the petroleum industry today, price is more or less given—that is, within a certain range, price is determined by the cartel, the Organization of Petroleum Exporting Countries. Hence, a firm can determine its total revenues by varying output, and the variation in output would depend on costs, given the price. In more technical terms, the elasticity we are interested in would be, not the elasticity of output in response to price changes, but the elasticity of output in response to costs, in so far as these affect returns. Included in these costs variables are government 'take'. A government analysing the effects of policies on discoveries may note that price increases provide a necessary condition but not a sufficient condition for firms to engage in new exploratory activity.

All studies of the oil industry recognize the huge capital outlay that must be committed before the first barrel of crude oil can be sent to the refinery. These include drilling, equipping fields, and building facilities to process, transport, and store crude. Hence, in any one year the flow of investment funds for a dynamic oil company is large compared to operating costs, or variable costs. Investment costs required to bring about production of petroleum resources may be grouped for convenience into two classes: exploration and development costs.

Exploration costs (also sometimes referred to as the cost of locating new reserves) include those elements involved in determining the location of hydrocarbons in preparation for drilling development wells and initiating production. As Figure 3.1 indicated, exploration activities generally begin with geophysical and geological surveys, and conclude with the drilling of exploratory wells. In general, geological and geophysical surveys are sunk costs in terms of investment decision, that is, these costs are not capitalized but are deducted from income in the year of expenditure.[5] Under certain types of petroleum development contracts, however, such as the Indonesian production-sharing contract, such expenditures for a particular contract area (except bonuses and the cost of capital) may be recovered from revenues, assuming that such expenditures for that particular area can be identified.[6] In theory this may be possible, but in practice it may be difficult to allocate parts of geological and geophysical costs to a particular area, if a firm's studies span several areas in a region.

In the drilling stage, the cost of exploration is a function of the cost of each exploratory well and the number of wells which are drilled on any given structure or tract. The cost of each well is a function of the overall environment (including drilling difficulty) and depth. The number of wells required to explore a structure varies significantly among structures.[7]

In the sense that the search for reserves provides knowledge concerning an area's potential for producing oil, exploration is analogous to research and development in the manufacturing industry. The information it generates—beginning with geological and geophysical studies up to exploratory drilling—determines whether or not additional capital should be invested in the next stage of the supply process. (See Figure 3.2 for a graphic distinction among the various types of wells leading to extraction.)

Development costs involve installing production wells and all facilities related to initiating production activity, and abandoning a depleted field. For onshore operations, they are a function primarily of drilling depth, dry-hole risk factors, drilling difficulty, completion costs, labour costs, and environmental conditions. In offshore production, platform costs, water depth, climate, and distance to the shoreline are additional variables.[8]

Development implies a new outlay of capital to drill more wells to delineate a reservoir or a field. The distinction between exploration and development can be very fine, and even blurred. For example, step-out wells or extension tests have all the characteristics of an exploratory well (see Figure 3.2 again), but may not correctly be capitalized as a development expenditure until the reservoir has been fully delineated. At the producing stage, the costs of labour and services related to extraction form part of operating costs or current expense.

A point of departure between the conventional cost-curve analysis and analysis of the cost of crude oil is the lack of parallelism between additional plant/equipment and additional wells. Additional wells in a reservoir may increase productive capacity, but such larger investment may not necessarily increase the total, cumulative production, unless secondary or tertiary methods are employed. Two variables determine total cost of production: the rate of output and the total volume to be produced. While allowing the crude in a reservoir to be produced sooner, increasing the number of wells also increases the investment outlays. Thus the time horizon for total production may be shortened but only at possible greater unit costs.

Other cost variables. Besides the physical cost elements related to investments in the project, a petroleum development firm must allocate some of its expected income stream for three types of items: (1) bonuses—whether they be signature or production bonuses; (2) taxes due to the host government or governments; (3) 'royal-

46

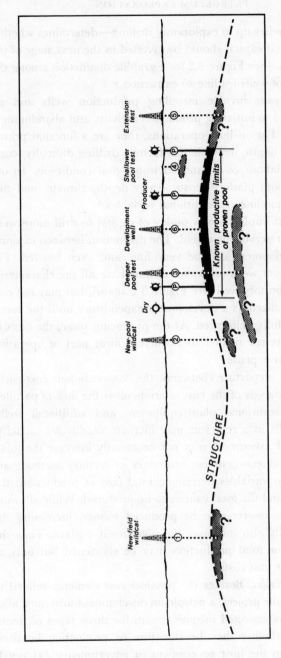

Figure 3.2. The Classification of Wells by Geologists

Proved reserves are established by the producing wells (P). Unproved reserves in the field will require additional drilling by wells 3, 4, 5 and 6. Wildcat exploratory drilling can find undiscovered resources in adjacent pools or in separate fields (1 and 2).

Source: Resources for the Future, *Resources*, No. 58 (March 1978), p. 4.

ties', in whatever form, e.g., share of production turned over to the government;[9] (4) cost of capital (that is, interest payments on any funds borrowed, or interest foregone on internally generated funds); and (5) external cost factors (such as fiscal policies in the petroleum firm's home country). The first three types of cost are referred to in this study as 'government "take"'.

THE NET REVENUE FUNCTION

The cost function for petroleum reserves may not be viewed independently of the anticipated income stream from a discovery field. Investments are made in the expectation of some size of revenue stream, and a producer will want to recover his investments over a certain period of time.[10] The period over which a reservoir is produced is determined, ultimately, by the cost/price relationship—at the point where net revenue is negative, the reservoir is abandoned. The unit cost of crude must at least reflect investment costs.[11] Since total operating costs increase by some factor through time, but remain constant in any time period regardless of the decline rate, unit costs increase at an exponential rate as production declines through time.[12]

The estimated revenues of a firm are not determined merely by the price of the discovered resource multiplied by the size of the discovery but are measured by the present value of the reserves. Since petroleum is produced from reservoirs over a period of several years, the value of a reservoir to a firm is the present value of the stream of output minus the present value of the costs associated with developing and producing an oil or gas field.[13] The net present value must be equal to the firm's alternatives—e.g., the maximum amount a firm would pay in order to obtain an identical reservoir of petroleum and the minimum amount it would be willing to accept if it sold the reservoir.

NON-COST VARIABLES

There are, of course, non-cost variables that enter the supply function. Technology might be subsumed under the various physical cost elements, but other factors that are external to the project are part of the decision to supply petroleum reserves. These include the presence of other more attractive alternatives outside the area

48

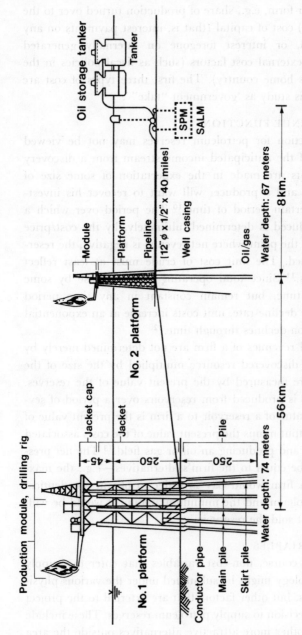

Figure 3.3. Diagram of Exxon Malaysia Platforms 250 Kilometres off Trengganu, Malaysia

Two offshore platforms constructed 250 km off Kuala Trengganu, Malaysia, are self-contained types having oil drilling and production facilities. A pipeline linking the platforms with a single-point mooring buoy runs a distance of 64 km, and also connects with a nearby oil storage tanker.

Oil produced and primary-processed on the platforms is delivered to an oil storage tanker (70,000–100,000 DWT). Then it is shipped to consumer countries by ocean-going tankers at intervals of seven to ten days.

Source: Nippon Steel, Japan.

for companies with multiple operations; or constraints on allocable investment funds (e.g., cash flow problems to fund multiple projects simultaneously, resulting in the ranking of such projects).

Another variable is the notion of risk. The search for oil and gas is, as already stated, a high-risk venture. Fisher stated: '... economic incentives not only influence the amount of exploration that occurs; they also determine its characteristics. Thus, ... an increase in economic incentives does lead to more wildcat drilling but ... this takes place on prospects poorer than those which can be drilled at a lower incentive level.'[14] Firms are likely to be risk averse, accepting an additional unit of risk only when they foresee something in excess of an additional unit of return.

The petroleum firm's profit function may ordinarily be expressed in the following form:[15]

$$\pi = \int [R(t) - C(t)] \, e^{-rt} dt \qquad (3.6)$$
$$R(t) = R \, (P, Q) \qquad (3.7)$$
$$C(t) = C(T, M, V, K, G) \qquad (3.8)$$
$$r = r(X_1, X_2) \qquad (3.9)$$

where R = revenue, C = costs, P = price, Q = output, T = technical costs (drilling and associated costs), M = mineral resource geological costs, V = variable costs (such as labour, drilling mud, etc.), K = capital costs, G = taxes, bonuses, production-sharing, and royalties, X_1 = geological risks, X_2 = economic and technical risks, r = the discount rate.

Increasingly, however, a third element—uncertainty over the future of the institutional framework—appears to be an essential feature of this function. We may call this X_3, or the 'political risk' element, so that:

$$r^* = r(X_1, X_2, X_3) \qquad (3.10)$$

The firm's perception of the magnitude of the political risk related to an investment may result in underinvestment in exploration and development if the firm discounts future revenues too heavily.[16]

Petroleum Exploration in South-East Asia

Interest in the petroleum resources of South-East Asia dates back to the late nineteenth century. The first wells were drilled in

Indonesia and Burma in 1872 and 1877, respectively. Exploration in the Philippines initially began in 1896. The Miri field in Borneo, now part of the Federation of Malaysia, was discovered in 1911, with exploration work in Brunei starting in 1910.

The structural changes of central focus in this study, however, occurred after the Second World War. One industry analyst divides the post-World War II period into four phases, in terms of level of activity. In 1976, Hatley wrote: 'Since the end of World War II, the exploration for and development of oil and natural gas have passed through three ... major phases of activity in East Asia. We are presently in a fourth ... phase of activity, which began in 1974.'[17] Hatley divides the post-World War II era as follows: Phase I—1946–63, the post-war 'rehabilitation period'; Phase II—1963–7, the stagnation period; Phase III—1967–74, the 'quiet boom'; and Phase IV—1974–?, the period of major reductions in activity. It is not clear from the evidence at this writing that Phase IV has actually ended and that a new era of renewed enthusiasm and vigorous exploration has begun.

Since our interest in the industry is not in its historical development as such, but in the economic implications of contractual arrangements and industry structure on the supply of petroleum reserves, the discussion in this section will be organized along the factors that led to the 'quiet boom' of Phase III and the 'downturn' since 1974 (Phase IV). In a broad sense, these factors—which were external as well as internal to both the region and to the industry—were essentially economic in nature, that is, cost/price relationships in the exploration for and development of petroleum resources had changed. The underlying reasons were, however, technological, political, and social in character.

TECHNOLOGICAL DEVELOPMENTS IN THE INDUSTRY AND COST/PRICE RELATIONSHIPS

The speculative nature of petroleum exploration has required sophisticated technology (1) in improving attempts at identifying underground sites of probable accumulations of petroleum, (2) in probing for such potential accumulations, and (3) in extracting discovered accumulations. Most of the land portions of the East Asian nations have been considered geologically unattractive, and until advances in offshore technology made searches in the South-East Asian seas possible, the region was all but ignored. The exception was Indonesia. Western rivalries in that country, in fact, began as early as the 1900s.[18]

Studies by oil companies as well as by non-profit bodies had, however, indicated very early the potential for petroleum accumulation in the offshore areas of the region. The studies most publicized are perhaps those of the UNDP/CCOP, the first of which was released in the organization's *Technical Bulletin*, Volume 2, issued in May 1969. The major oil companies had, however, already begun conducting their own studies as early as the 1950s[19] and 1960s.[20]

The costly nature of offshore technology, however, requires the appropriate investment 'climate', and the East Asian region has had to compete with other regions in attracting such investments. The highly capital-intensive nature of oil exploration, but even more so of offshore exploration, and the prevailing cost/price relationships in the 1950s and early 1960s limited such searches in that period to the major oil companies. Following the years of reconstruction, East Asia's potential as an investment site remained relatively less attractive than other areas for several reasons. Among these reasons were (1) major discoveries in Africa, the North Sea, and Latin America, (2) a prevailing surplus production worldwide, and (3) the resulting decline of crude oil prices in actual and real terms.[21] These provided very little incentive to the major companies—who were faced with more attractive alternatives—to allocate a significant portion of their internally generated funds to South-East Asian exploration.

Several factors may have served to renew interest in applying advanced offshore technology in East Asia's waters: (1) the offshore discovery in Brunei in 1963 by Shell, and (2) efforts of the Indonesian government to attract smaller oil companies with the novel 'production-sharing' contract.

Continued advances in the industry's offshore know-how and equipment, reassessments of the region's potential with such techniques—given other political and economic factors affecting

the industry world-wide—further enhanced the investment attrac-
tiveness of the region.

GLOBAL CHANGES IN THE OIL INDUSTRY

By the late 1950s and certainly in the 1960s, the role of government policies in shaping the oil industry became more significant
in producing as well as in consuming countries. In the developing
countries, in particular, governments became increasingly aware of
the role the industry could play in achieving their economic
development goals, and therefore more concerned about international prices, balance of payments, the lives of their reserves, the
location, ownership or control of producing and refining facilities,
as well as foreign investments.

Two features stand out about the structure of the international
oil industry: (1) the vast size of the companies and (2) the high
degree of vertical integration commonly found. Seven companies
(commonly referred to as the 'majors') dominate the international
market. The typical company is active at all levels of the
industry—from exploration and production to transportation and
refining. This was true in South-East Asia until the 1960s when
smaller and what are often termed 'independent' oil companies
entered the exploration scene.[22]

The emergence of vertically integrated firms was by no means
the result, initially, of a technical or economic 'necessity', as Penrose points out,[23] but largely the result of government policies that
permitted such growth, both in the U.S. (where most of the majors
are based) and abroad. The terms at which concessions were
obtained in the producing countries, as well as the size of the
financial resources which the international companies had
attained, provided the companies with significant discretion in
producing, pricing, and marketing their crude oil.

A turning point in the world-wide position of the major oil companies vis-à-vis host countries with petroleum resources was the
establishment of OPEC in 1960. The erosion of international crude
oil prices after 1958 and vigorous competition among the sellers set
the stage for a fundamental and irreversible change in the position
of the oil companies relative to the host governments. Increasing

competitive pressures in the world market after 1958 (especially in Europe and Japan) led the oil companies to reduce both posted[24] and realized prices by 8 per cent in 1959 and by 3 per cent in 1960 (see Table 3.1).

OPEC was formed largely in response to these price cuts, and one of their first stated objectives was to restore posted prices to 1958 levels. (This was not achieved but posted prices stabilized between 1961 and 1970; see Table 3.1 again.) At this point also, the host governments had become aware of the value of the resources they held—to the companies and to the world in general. They had also advanced in their political, economic, and administrative expertise.[25]

The result was a drastic alteration in the property rights arrangements between the oil companies and the host governments in the OPEC countries, and eventually in other producing, developing countries. Oil pricing policies in particular, but resource development policies in general, are no longer solely, nor necessarily, determined by simple economic considerations but by a complex multitude of factors within the framework of each individual government's multiple objectives.

STRUCTURAL CHANGES IN SOUTH-EAST ASIA

Developments in the petroleum industry in South-East Asian countries may not be viewed except in the context of what happened elsewhere. Awareness by a host developing country that the resources it owned and which were desired by foreign firms were a source of bargaining strength at the negotiating table changed relationships between governments and foreign investors even in Asia. As the next section will show, Law No. 44 in Indonesia was passed in 1960, Indonesia joined OPEC in 1962,[26] and negotiations on new contractual arrangements culminating in the first production-sharing contract began in 1963. Later, Malaysia passed its Petroleum Development Act of 1974 which terminated all concessions issued under previous legislation and required renegotiation of terms under which its petroleum resources could be produced, once discovered. Other countries, like the Philippines and Burma, also altered their contractual frameworks.[27]

TABLE 3.1

PETROLEUM PRICES: POSTED AND REALIZED, 1950–1976
(US$/barrel)

Year	Saudi Arabian (Posted Price) [1]		Saudi Arabian (Realized Price) [2]	
	Current ($)	1975 Constant ($)	Current ($)	1975 Constant ($)
1950	1.71	4.70	1.71	4.70
1951	1.71	3.91	1.71	3.91
1952	1.71	3.85	1.71	3.85
1953	1.84	4.32	1.84	4.32
1954	1.93	4.63	1.93	4.63
1955	1.93	4.56	1.93	4.56
1956	1.93	4.43	1.93	4.43
1957	2.01	4.47	1.86	4.13
1958	2.08	4.70	1.83	4.13
1959	1.92	4.39	1.56	3.57
1960	1.86	4.15	1.50	3.35
1961	1.80	3.98	1.45	3.21
1962	1.80	3.99	1.42	3.15
1963	1.80	3.97	1.40	3.08
1964	1.80	3.94	1.33	2.91
1965	1.80	3.88	1.33	2.87
1966	1.80	3.79	1.33	2.80
1967	1.80	3.76	1.33	2.78
1968	1.80	3.78	1.30	2.73
1969	1.80	3.64	1.28	2.59
1970	1.80	3.41	1.30	2.46
1971	2.21	3.94	1.65	2.94
1972	2.47	4.05	1.90	3.12
1973	3.30	4.56	2.70	3.73
1974	11.59	13.10	9.78	11.05
1975	11.53	11.53	10.72	10.72
1976	12.38	12.20	11.51	11.34

Source: World Bank, *Commodity Trade and Price Trends*, 1977 Edition (August 1977), Report No. EC-166/77.

[1] Light crude oil, 34°–34.9° API gravity, f.o.b. Ras Tanura.

[2] Light crude oil, 34°–34.9° API gravity, average realized price, f.o.b. Ras Tanura.

Note: See Appendix A for the definition of API gravity.

The wave of 'nationalistic' laws and policies which swept through South-East Asia in the mid-1960s was new to the major oil companies, but also coming in the wake of other alternatives, helped to dampen investment initially. This was partly offset by technological developments (discussed earlier) and later by other developments in the industry related to the acceptance of the concept of a production-sharing formula in production. Discoveries in Thailand, Malaysia, and Vietnam and increases in the 1970s in crude oil prices, both in current and real terms, provided further impetus that led to the 'quiet boom' which reached its peak in 1974.

In the meantime, however, the structure of the exploration segment of the industry shifted from one where the investors were a few giants to a mixture of major, minor, and independent oil firms. Investments also tended to be joint venture arrangements among two or more companies, not necessarily 'independents'. In the Indonesian case, one of the partners would be the party to a production-sharing contract with the national company and would be the legally recognized contractor,[28] even where more than one company holds an interest in the contract area.

OTHER FACTORS AFFECTING PETROLEUM RESERVE SUPPLY

There are factors affecting the level of exploration and development of South-East Asia's petroleum potential and the supply of petroleum reserves which are not directly connected with investment incentives. It is not the intention of this study to dwell on them, but at this juncture it may be useful to point out two specific cases.

A clear case is Burma's onshore resources. The Burmese government has reserved to the state the task of developing its onshore resources, and, therefore, foreign incentives are not relevant. Even if the state government had the capital resources to expand the country's onshore petroleum reserve supply, however—which reports indicate it does not have—it is not able to explore for and develop Burma's full onshore potential because certain areas containing petroleum accumulations are outside the control of the central government. The Arakan region on the west coast (where there are some small producing wells) has been the subject of secessionist threats, while most of the northern part of the country is held by insurgents.[29]

Another case is that of Vietnam. The exploration wells off the coast of Vietnam drilled in 1974 and 1975 yielded two discoveries out of six wells drilled.[30] Two promising wells were not completed because of the heated political conditions between the two Vietnamese governments in September 1974.[31] Active offshore exploration since reunification of the North and the South resumed only in 1977 but progress was slow. This was due partly to the inability of the present government to obtain sufficient suppliers of offshore technology.[32] Another factor was the ban on U.S. company involvement placed by the U.S. State Department following North Vietnam's take-over of the South.[33] This not only kept the American companies from resuming exploration but created uncertainty over property rights acquired from the former South Vietnamese régime, especially since significant signature bonuses had been paid to the latter.[34] The first offshore well following the suspension of offshore activity in 1975 was reportedly drilled 200 miles south of Ho Chi Minh City (formerly Saigon) in the South China Sea.[35] This well was dry; the second well drilled by the same company was also dry. Two other contractors were scheduled to drill at this writing.[36]

The Indonesian Contractual Framework: A Brief Discussion of Its Development

The Indonesian production-sharing contract has been hailed as an innovation in the industry and has set the pace for exploration activity and government legislation in the region. For that reason, special attention will be paid to its development and the important events surrounding it.

THE CONSTITUTION AND INDONESIAN PHILOSOPHY

Although Indonesia proclaimed its independence in 1945, the Mining Law of 1899 under the Dutch colonial régime remained

in force until the enactment of the Petroleum Act of 1960. The
Mining Law of 1899 had provided for the separation of surface
versus subsoil rights. Rights to develop subsoil petroleum resources
were reserved to the government which could then contract out
such rights to private parties in the form of concessions.[37]

The 1945 Constitution of Indonesia reaffirmed state control over
its natural resources.[38] Law No. 44 of 1960, citing paragraphs 2
and 3 of Article 33 of the Constitution, and singling out the unique
role of oil and gas to achieve the development goals of the nation,
set out the rules governing oil and gas extraction in Indonesia.

CONTRACTS OF WORK (*PERJANJIAN KARYA*)

Law No. 44 specifies that the mining of oil and gas in Indonesia
would be undertaken by the state and carried out solely by a state
enterprise (Article 3). Other parties could, however, engage in such
mining as contractors to the state enterprise (Article 6). Holders of
concessions[39] under legislation existing prior to the adoption of
Law No. 44 were allowed to continue under the terms of their concession for a 'period of time which shall be as short as possible...
determined by government regulation' (Article 22, Section 1). During this 'grace period' new arrangements were negotiated. With the
conclusion in September 1963 of what are called 'contracts of work'
or *perjanjian karya* (sometimes also referred to as *kontrak karya*), the
major companies concerned—Stanvac, Caltex, and Shell—
relinquished their existing concessions to the government and
began operating under the new terms (although Shell eventually
sold all its interests to the Indonesian government in 1965).[40] A
contract containing similar terms was earlier signed on 15 June
1962, with Pan American International Oil Corporation, a subsidiary of Standard Oil Company of Indiana,[41] but was terminated
on 3 October 1966.[42]

In terms of duration, there were two types of contracts: (1) that
covering already producing areas (sometimes referred to as 'old
areas'), which was to run for 20 years; and (2) that covering 'new
areas', which was to run for 30 years (10 years for exploration and
20 years for production). The Caltex contract was of the first type,
and the Stanvac contracts were of both types.

In the contract of work, the features of which are outlined in greater detail in the next chapter, management control remained with the contractor. The contract also provided for the sale to the Indonesian government of all refineries and marketing assets over a period of time. For refineries, this was to be done beginning 10 years from the date of the contract, with final turnover by the fifteenth year; for marketing facilities, this was to begin within 5 years. The firms were also to provide the domestic market with petroleum at cost plus fees. The profit split was at 60/40 in favour of the Indonesian government, and the state could elect to take 20 per cent of aggregate production in kind.[43] The latter provision assured the Indonesian government a minimum income of 20 per cent of aggregate production.

Caltex has negotiated several production-sharing contracts to take care of operations on acreage it holds, upon expiration of its work contracts; one was negotiated in 1971, another in 1975.[44] Another was reported to have been negotiated in 1978.[45]

THE PRODUCTION-SHARING CONTRACTS

Pertamina (which stands for Perusahaan Pertambangan Minyak dan Gas Bumi Negara) dates the signature of the first production-sharing contracts (PSCs) with those concluded with Refining Associates of Canada, Ltd. (Refican) on 10 June 1961 (amended in 1964 to include more precise legal terms) and with Asamera Oil Corporation, Ltd. on 1 September 1961.[46] These contracts contained the management clause that is considered one of the main differences between the concession and the PSC. The contracts also had the two other main features of the PSC, production-sharing and cost recovery clauses. Refican's contract provided for a division of output from rehabilitated wells at 65/35 in favour of the government and from new discoveries at 60/40, with a 40 per cent cost recovery for materials and equipment.[47] Asamera's contract had a similar format.

The contract whose format and conditions became the basis for future production-sharing contracts was, however, that signed between the Indonesian government (through its state enterprise Permina)[48] and the Independent Indonesian American Petroleum

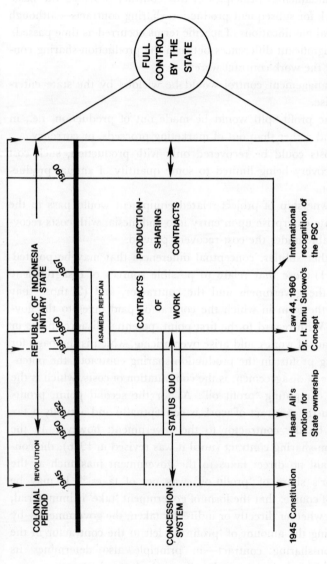

Figure 3.4. Diagram showing the Developments towards National Control in the Indonesian Petroleum Industry

Source: Pertamina.

Company (IIAPCO) on 18 August 1966.

The fundamental principles of this contract provided the basic framework for subsequent production-sharing contracts—although substantial modifications of specific terms occurred as time passed.

Four significant differences between the production-sharing contract and the work contract were:

(1) Management control would be wielded by the state enterprise.

(2) The profit split would be made out of production, i.e., in kind rather than out of marketing proceeds, or currency.

(3) Costs could be recovered only with production, such cost recovery being limited to some quantity of annual production.

(4) Ownership of project-related equipment would pass to the state enterprise upon entry into Indonesia, with costs recoverable under the cost recovery proviso.

Two other subtler, conceptual differences that may be pointed out are (1) those that relate to possible sources of disagreement between the government and the contractor, and (2) those that relate to the form in which the contractor paid taxes to the government. With regard to the first point, potential disagreements in the contract of work could arise over pricing, which is the basis for computing profits; in the production-sharing contracts, the potential source of disagreement is the computation of costs, which is the basis for computing 'profit oil'. As for the second point, profit-sharing in the contract of work is conceptually and actually a tax payment by the contractor to the government; however, in the production-sharing contract (until it was revised in 1976), the contractor paid no direct taxes to the government inasmuch as the contractor's share of 'profit oil' was net of taxes.[49] It may be argued, of course, that the form of government 'take' is immaterial, and that, whether directly or indirectly taken, the government—by determining the amount of 'profit oil' left to the contractor in the production-sharing contract—in principle also determines its revenues. This is borne out by the method of calculating government 'take' in the 1976 revisions of the PSC (see Chapter IV).

An additional difference that might be noted arises from the computation of the pro rata domestic allocation of crude output. In the *perjanjian karya* the company's share is computed off its total output ('off the top'); thus the company is reimbursed for costs plus fee (20 U.S. cents at this writing).[50] In the PSC, the company's share is computed off its after-cost share of production in its contract area; hence the company is only paid the fee of 20 U.S. cents.[51] (This fee was modified for a temporary period beginning in 1977; see discussion below on new incentives in 1977.)

MODIFICATION OF CONTRACTUAL TERMS

The general terms of the production-sharing contract have remained the same over the years, but the specific provisions on production-sharing proportions and cost recovery have changed (see Chapter IV). The standard production-sharing contract originally provided that the crude remaining after allowance for cost recovery would be split 65 : 35 in favour of the state enterprise. Contracts signed from mid-1967 onwards have provided for a graduated increase in the share of the state upon reaching a certain production level during any previously agreed period, at levels ranging from 50,000 to 500,000 barrels per day (see, for example, Continental Oil Company's 1967 contract and INCA Limited's 1976 contract).[52] The basic profit split also changed.

With the dramatic increase in international oil prices in 1974, the Indonesian government insisted that the resulting windfall profits to oil companies should largely benefit the state. Two alterations were made in the *perjanjian karya* of Caltex—which served as the forerunners of subsequent changes in other existing agreements as well as in new production-sharing contracts. The alterations, which were made public in February 1974 but were retroactive to 1 January 1974, involved the division of 'profit oil' into tiers and the use of 'base price' for computing increasing levels of the government share.

Under this arrangement, the 60/40 split of net income on the base revenue was retained; this base was initially fixed at US$5.00 but was subject to escalation according to a formula using the UN price index for manufactured exports.[53] By August 1976, this base was reported to be US$5.93.[54] The split was also altered upwards on a sliding scale: 85/15 for the first 150,000 b/d production, 90/10 on the next 100,000 b/d and 95/15 on output levels above 250,000 b/d.[55]

Production-sharing contracts negotiated in 1975 reflected this move to increase government 'take', on the one hand, and on the other hand, it probably indicated the degree of interest of contractors in certain contract areas despite less favourable terms. The Phillips-Tenneco contract of February 1975, for example, provided for a maximum of 35 per cent cost recovery, a 72.5/27.5 'profit oil' split rising to 77.5/22.5 at 50,000 barrels per day, and to 80/20 at 150,000 barrels per day. The signature bonus was $3 million, and the production bonus $2 million; an earlier contract with Agip reportedly required $1.5 and $0.5 million. Expenditure commitments were also significantly steeper per square kilometre.[56] (The financial conditions in Appendix B, which were negotiated earlier, are far steeper than those mentioned.)

THE RENEGOTIATIONS OF 1975/1976

To raise additional revenues to cope with the 1975 Pertamina crisis,[57] existing contracts were renegotiated in late 1975 and in 1976. The Indonesian government started talks with Caltex in October 1975 on an additional 'surcharge' of $1.00 for every barrel of oil produced in their *perjanjian karya* areas, with payments to begin in January 1976. In April 1976 the Minister of Mines disclosed that Caltex had agreed to this; Stanvac—the other producer under a contract of work—also signed the contract in 1976.[58]

In January 1976, the Indonesian government also announced it would revise the cost-recovery formula and profit-split in existing production-sharing contracts. Failure to obtain quick concurrence on the part of oil companies resulted in a deadline (originally 31 July 1976, and later extended to 14 August 1976) for signing the revised PSCs. The companies eventually agreed to the new terms.

The companies were divided into two groups. In Group I were those whose proven reserves could produce for 7 or less years; in Group II were those whose proven reserves had an estimated remaining production life of more than 7 years. Instead of the earlier 40 per cent cost recovery 'off the top', the double-declining

TABLE 3.2
BASE PRICE ADJUSTMENT

The new quarterly Base Price is the lesser of:

1. $\$5 \times \dfrac{\text{UN Price Index last previous cal. } \frac{1}{4}}{\text{UN Price Index first cal. } \frac{1}{4} \text{ 1974}}$ = New Base Price

2. $\$5 \times \dfrac{\text{Wt. Av. Value last previous cal. } \frac{1}{4}}{\$10.80}$ = New Base Price

Assume that in the first calendar quarter of 1974 the UN Price Index for manufactured goods exports is 100 and that the weighted average value of a barrel of Crude Oil is $10.80. During the second calendar quarter of 1975 the said figures rise to 110 and US$11.00 respectively.

1. $5 \times \dfrac{110}{100} = 5.50$

2. $5 \times \dfrac{11.0}{10.80} = 5.09$

The new Base Price effective 1 July 1975 would be US$5.09.

Source: Pertamina.
Note: 'Cal. ¼' means 'calendar quarter'.

depreciation method for capital expenditures was used. Depreciation for Group I companies was to be spread over a 7-year period; for Group II companies this would be done over a 14-year period. Non-capital costs for both Group I and Group II companies could be expensed without limit in the year of expenditure but those carried forward would be recovered on a straight-line basis at 8 per cent interest. The balance—or 'profit oil'—would then be split 85/15 in favour of the state.[59]

Allocation for the domestic market remained at cost plus fee for contract-of-work holders and at the fee for PSC holders.[60]

THE IRS RULING AND THE PRODUCTION-SHARING CONTRACTS

The mechanics of the production-sharing contracts created difficulties for the Indonesian government and the contracting companies when the U.S. Internal Revenue Service (IRS) issued Ruling No. 76–215 on 7 May 1976. The IRS disallowed a foreign tax credit to a U.S. oil company for the share of production retained by the state. The IRS ruled that such crude oil paid to the government was a royalty in its entirety and was therefore not eligible for a foreign tax credit in the United States.

The arguments on which the ruling was based were:

(1) the production sharing contract provided the sole source of revenue for the Indonesian Government and the retained share of production was the Government's only compensation for the exhaustion of oil deposits to which the Government had title; (2) the recovery of signature and production bonuses and interest paid on borrowed money used for petroleum operations was not allowed and no recovery of annual operating costs in excess of 40 percent of the value of all barrels of oil produced and saved from the contract area during the year was allowed; (3) the income from each production-sharing contract was computed separately from the income under other production-sharing contracts held by the same party and a loss under one contract could not be offset against income earned under other contracts; and (4) the Indonesian Government was assured a share of production regardless of whether income has been realized.[61]

This was followed by an announcement by the IRS on 14 July 1976 detailing the bases for tax credits on payments to foreign governments in connection with mineral extraction. The following were considered requisite:[62]

(1) The amount of income tax is calculated separately and independently of the amount of the royalty and of any other tax or charge imposed by the foreign government.

(2) Under the foreign law and in its actual administration, the income tax is imposed on the receipt of income by the taxpayer and such income is determined on the basis of arm's length amounts. Further, these receipts are actually realized in a manner consistent with U.S. income taxation principles.

(3) The taxpayer's income tax liability cannot be discharged from property owned by the foreign government.

(4) The foreign income tax liability, if any, is computed on the basis of the taxpayer's entire extractive operations within the foreign country.

(5) While the foreign tax base need not be identical or nearly identical to the U.S. tax base, the taxpayer, in computing the income subject to the foreign income tax, is allowed to deduct, without limitation,

the significant expenses paid or incurred by the taxpayer. Reasonable limitations on the recovery of capital expenditures are acceptable.

The foregoing requirements had two serious implications. First, the Indonesian government had to redefine its cost recovery principles and, second, its tax collection method would have to be modified if U.S. companies were to be able to claim U.S. tax credits for income tax payments to the Indonesian government. Under the original production-sharing scheme, only exploration costs incurred in an area with commercial production were recoverable; all others were completely lost. Also, income tax payments to the Indonesian government were made by the state enterprise on behalf of a contractor, such payments being made out of the state enterprise's share of 'profit oil'.

The second implication was less difficult to resolve than the first one (on recovery of pre-production costs). Under the 1976 revisions, a new system was adopted whereby the contractor would pay a portion of the 'profit oil' split directly to the Indonesian government rather than to the state oil company. Thus the new contracts provided for the following kind of split. After deducting costs, a corporate tax is levied on the balance or 'profit oil' before this is split 65.91/34.09 in favour of the Indonesian government. The corporate tax is 45 per cent of 'profit oil'. A dividend tax—with an effective rate of 11 per cent—computed against the 34.09 per cent share minus 45 per cent corporate tax, is also levied.[63] This formula results in a final effective 85/15 split of the profit crude.[64]

As for the first implication, a change that would allow an oil contractor to recover *all* current exploration costs in Indonesia from current production in a contract area would (1) reduce the total amount of oil to be divided between Indonesia and the contractor, and (2) would tacitly force Indonesia to share in the risk of exploration. For small fields, it is possible to have a situation where the Indonesian government would receive no revenues in the earlier phase of operations in a contract area.[65] The issue arises especially if exploration is in progress in one or more blocks falling under the same contract that simultaneously covers a producing block. This means that intangible drilling costs in one block may be recovered currently from another block that is already producing; should that exploration block turn out to be totally worthless, total output share of the government from the contract area would be reduced by the procedure.

At the time of this writing, differences were reported to have been removed in principle, and consolidation of costs and profits over the whole contract area had been agreed upon by the government and by the foreign contractors. In practice, it was expected that the following could result:

(a) Consolidation would be agreed upon with regard to the old contracts. That is, if there are two or more blocks under the same contract, the costs incurred in the unsuccessful block or blocks would be recoverable from production in the successful block or blocks.

(b) The Indonesian government will not share the risk of exploration for new contracts, and so separate companies would have to be formed for each new contract area. That is, companies would establish separate legal entities for each new contract area.[66]

NEW INCENTIVES INTRODUCED IN 1977

Two sets of incentives were introduced in 1977 to lift exploration activity out of the depressed state of 1976. These incentives were *de facto* reductions in government 'take' on companies, especially those producing 'new oil' (defined as oil coming from fields brought into production in 1977 or later as well as from new development of existing fields through secondary recovery activities) or making certain types of new investment.[67]

The *first set* of investments was announced on 11 February 1977, and had three main elements:[68]

(1) Oil companies would be exempt for five years from providing Pertamina with a share of their 'profit oil' for domestic use at cost plus 20 cents per barrel. Instead, for five years companies would provide the fuel at the 'prevailing price' (defined as the full market price for Indonesia). Allowing the extra payment is 'based on the understanding that the additional proceeds' would be used 'to assist financing of continued exploration in the company's contract area', according to a letter reportedly sent to the companies.

If no such opportunity could be demonstrated to exist, the contractor would be allowed to use the proceeds as it saw fit.[69]

(2) The relevant companies would receive a 20 per cent credit for capital investments 'directly- required for developing production facilities of new oil fields' in areas where start-up costs are high. These 'difficult areas' were defined as the onshore regions more than 30 miles from the coast or the nearest oil terminal or pipeline, and offshore areas with a depth of more than 300 feet. The investment credit would be taken up by the oil companies through a deduction from gross production in the first year of new production, i.e., it may be taken before the producers turn over 85 per cent of their 'profit oil' to Indonesia under the terms of the production-sharing contract.[70]

(3) Capital investments related to this 'new' production could be recovered over seven years, using a double-declining balance depreciation system.

The *second set* of incentives came in a form that might be termed 'rebates' for 'new' oil produced by contract-of-work holders, i.e., Caltex and Stanvac. Beginning 1 January 1977, such companies could withhold 50 cents per barrel from new fields or from existing fields where secondary recovery methods were employed.[71] Pertamina and Caltex were reported to have signed a five-year agreement on 18 May 1977. Stanvac had no secondary recovery projects as of May 1978, and no agreement was, therefore, signed.[72]

PERTAMINA'S JOINT VENTURES

In early March 1977, Pertamina announced its intentions of seeking foreign participation in the exploration and development of areas designated for its own exploitation. Under this plan, foreign companies would be allowed to explore and develop these areas, with costs and outputs shared on a 50-50 basis. The areas under consideration were located in Sumatra, Kalimantan, and Irian Jaya.[73]

On 14 April 1977, the press reported that foreign oil firms had been formally invited to submit tenders to participate in two blocks on the south coast of the Bird's Head Peninsula in Irian Jaya.[74] A deadline of 7 June 1977 was set for tenders, and in exchange for inspection fees, interested firms could examine data available on the blocks.[75] The details of the plan were outlined as follows:

(1) Pertamina and foreign firms would bear all costs of exploration, development and production from the jointly operated concession on a 50-50 basis. The foreign company would carry all expenses until Pertamina's previous expenditures in the block were matched, or until three consecutive years of exploration were completed, whichever came later.[76] (These were on areas where Pertamina had already incurred expenditures and did not want to invest more for the time being.)

(2) The parties would divide equally all production of gas/oil in the concession.

(3) They would conclude a standard production-sharing agreement on the 50 per cent portion retained by the foreign firm.

In May it was reported that Pertamina and Caltex had signed one of these agreements covering one of Pertamina's concession areas off Irian Jaya.[77] Agreements covering the two onshore blocks offered in Irian Jaya were signed with Total Indonesie and Conoco Irian Jaya Co. on 22 October 1977.[78] The contract provides, among other things, for 5 per cent participation in the companies' interests by Indonesian entrepreneurs should commercial production be achieved.[79]

In July 1977, a block in north-east Kalimantan was reportedly offered to oil companies,[80] and in March 1978, onshore areas in Central and South Sumatra were reportedly also offered.[81]

1. There are important reasons for wanting to distinguish the supply function of reserves (which we shall call S_r) from the supply function of production (which we shall call S_q). This will become clearer as the discussion progresses. Suffice it to say at this point that S_q may not be the same as S_r if other factors influence the rate of output such that the decline rate a in Equations 2.2 and 2.3 (in footnote 6, Chapter II) differs according to the investment setting.

2. The 'downstream' phase covers refining and marketing.

3. 'Risk' is distinguished from 'uncertainty' in the degree of the opportunity of loss; the term 'risk' refers to an opportunity for loss, whereas the term 'uncertainty' may be applied to 'factors where the outcome is not certain but where the opportu-

nity for loss is not as apparent as in risk'. This distinction is made in Megill, *Exploration Economics*, p. 98.

4. Rummel and Heenan distinguish between 'political uncertainty' and 'political risk'. They define *political uncertainty* as an unmeasured doubt about a political environment, while *political risk* is a term they use to denote 'a relatively *objective* measurement, usually resulting in a probability estimate of that doubt'. (See R. J. Rummel and David A. Heenan, 'How Multinationals Analyze Political Risk' in *Harvard Business Review*, Vol. 56 (Jan.–Feb. 1978), pp. 67–76.) For our purposes, we will not distinguish between the two concepts but will treat both political uncertainty and risk as 'political risk', in respect of how the foreign petroleum contractor views changes in petroleum policy, although the probability of such changes may not be measured over a long-term period.

5. Megill, op. cit., pp. 22–3.

6. A contractor is required to treat all costs incurred with respect to a contract area prior to the commencement of production as *pre-production costs*. Pre-production costs that relate to capital equipment are recoverable through depreciation deductions, beginning with the year in which such equipment is placed in service. Intangible drilling and development costs plus all other costs not related to capital equipment are recoverable through amortization deductions beginning with the year in which production is begun in that contract area. The amortization deductions are taken over a period equal in length to that over which a contractor would depreciate production facilities. (See summary of Rev. Ruling 76–215, Appendix C.)

7. See Paul Weaver, 'Variations in History of Continental Shelves', *Bulletin of the American Association of Petroleum Geologists*, Volume 34 (1950), cited in Kalter, p. 19.

8. See Figure 3.3 for a diagram of the Exxon platforms off the Malaysian peninsula.

9. The term is used here loosely to refer to the output share of the owner of the resource that is in no way related to cost of production but that accrues to the recipient by virtue of his ownership of the resource being developed. (See the more restrictive definition on page 103, note 9.)

10. A producer may be said to want to maximize total cumulative output, that is:

$$(3.1) \quad \max \int_0^T q_o e^{-at} \, dt = S_o$$

Equation 3.1 states that for a given production time horizon, the maximum cumulative output is equal to the magnitude of reserves which may be recovered at a given rate of production and installed capacity.

11. He may estimate his costs as follows:

Let θ = physical parameter related to initial reservoir conditions

K_o = initial operating costs per unit of capacity. Then

$$(3.2) \quad q_o K_o e^{\theta t} = \text{total operating costs at any point in time.}$$

Unit costs would then be:

$$(3.3) \quad q_o K_o e^{\theta t} / q_o e^{-at} = K_o e^{(\theta + a)t}$$

12. See equation (3.3).

13. Assuming certain engineering and geological conditions, and assuming some knowledge of the size of the reservoir, this may be represented as follows:

$$(3.4) \quad \frac{d}{dq} (R_o - K'_o) = \frac{d}{dq} (R_t - K'_t) \left[\frac{1}{(1+i)} t \right] (1 + s_t)$$

where

$$(3.5) \quad K' = K_o e^{(\theta + a)t}$$

$\dfrac{d}{dq} R$ = long-run marginal revenue

$\dfrac{d}{dq} K'$ = long-run marginal costs

$\quad i$ = the rate of interest for the producer, adjusted for risk and uncertainty

$\quad s_t$ = the fraction of a barrel lost from ultimate recovery and which might have been recovered in time t, for every barrel of production transferred from time t to time o; that is, for every barrel produced beyond the optimum rate.

14. F. Fisher, *Supply and Costs in the U.S. Petroleum Industry* (Baltimore: The Johns Hopkins University Press, for Resources for the Future, 1964).

15. We thank an anonymous referee for improving the formulation of these equations.

16. The risk-adjusted discount rate i in Equation 3.4 may be expressed as follows:

$$i = r + k^\sigma$$

where r = the risk-free discounted rate,

$\quad \sigma$ = the standard deviation of the payoff, which changes with the contractual terms.

$\quad k$ = a constant related to the degree of risk preference or aversion, and to uncertainty related to costs or 'political risks'.

17. Allen G. Hatley, 'Offshore Petroleum Exploration in East Asia—An Overview', Paper 1, SEAPEX Program, Offshore Southeast Asia Conference, February 1976.

18. See Edith T. Penrose, *The Large International Firm in Developing Countries* (London: George Allen and Unwin Ltd., 1968), pp. 54–5.

19. Shell started its first offshore exploration well in Brunei in 1954. See *PE*, June 1978, p. 248.

20. Esso had already done studies of areas in the Sulu Sea of the Philippines in the early 1960s. See, for example, Cora Siddayao, 'Looking for Oil in the Philippines', *Esso Eastern Review* (New York, June 1966).

21. See Hatley (1976), op. cit., p. 3.

22. The term 'independent' is generally used to refer to companies operating only in one phase of the industry, e.g., production or marketing.

23. See Penrose, op. cit., pp. 84–5.

24. Posted prices are tax-reference prices.

25. See Penrose, op. cit., pp. 198–200.

26. See Anderson G. Bartlett III *et al.*, *Pertamina* (Jakarta: Amerasian Ltd., 1972), p. 187.

27. See M. Rajaretnam, *Politics of Oil in the Philippines* (Singapore: Institute of Southeast Asian Studies, 1973), and Ng Shui Meng, *The Oil System in Southeast Asia* (Singapore: Institute of Southeast Asian Studies, 1974) for background discussions on developments in the region.

28. Based on discussion with Legal Division, Directorate General of Oil and Gas.

29. See *PE*, December 1978, pp. 505–6, 'Call for exploration bids'.

30. See American Association of Petroleum Geologists, *Bulletin*, 'Petroleum Developments in Far East in 1974', October 1975 issue, and 'Petroleum Developments in Far East in 1975', October 1976 issue.

31. For a discussion of the issues involved, see Siddayao (1978), op. cit., pp. 76–83.

32. For details, see *PE*, February 1979, pp. 62–3, 'Problems loom for renewed search', and *PN*, December 1978, Supplement.

33. This ban is based on the Trading with the Enemy Act of the U.S., implemented in 1975, and renewed annually since then. See *PE*, February 1979, p. 62.

34. Three European companies—Deminex (of West Germany), Elf Aquitaine (of France), and AGIP (of Italy)—were awarded blocks in 1977 that had been held by American companies under concession from the defunct South Vietnam government. (See *PN*, January 1978, p. 45.) The Elf negotiation fell through in 1978, according to subsequent reports (see *PE*, February 1979, p. 65), but these moves have, nevertheless, been interpreted as expropriation without compensation.

35. This was a well drilled by a group of Canadian oil companies. The contract for this area was signed in September 1977, and the 3.4 million acre concession is operated by Bow Valley Industries Ltd. Bow Valley held previous rights to this concession under an award made in July 1973 by the former South Vietnamese régime. See *AWSJ*, 14 February 1979, p. 2, 'Canada firms search for oil off Vietnam'.

36. See *PE*, April 1979, p. 174, and *PE*, May 1979, p. 214.

37. This is a summary of a discussion on the subject by Oei Hong Lan, 'Petroleum Resources and Economic Development: A Comparative Study of Mexico and Indonesia', Ph.D. dissertation, University of Texas, 1964 (Ann Arbor, Mich.: University Microfilms, Inc.).

38. As an Indonesian official explained it, the philosophy behind this was that the natural resources of the nation belonged to the people; and the state, as the trustee, would administer its exploitation in a manner that would redound to the benefit of the people.

39. These concessions were also sometimes referred to as '5A contracts' because they referred to Section 5A of the Mining Law of 1899, which stipulated the provisions for contracts between private enterprise and the government. This point was brought up by an Indonesian government official. See also Oei, op. cit., p. 94, for the explanation on the Mining Law. The terms 'concession' and '5A contracts' will be used interchangeably in the text.

40. See Sritua Arief, *The Indonesian Petroleum Industry: A Study of Resource Management in a Developing Economy* (Jakarta: Sritua Arief Associates, 1976), pp. 302–3. In this present study, the terms *perjanjian karya* and *kontrak karya* will be used interchangeably, although legalists may prefer the former.

41. Oei, op. cit., p. 115, citing Agreement text.

42. This information was provided by the Indonesian government.

43. This is summarized from Bartlett, op. cit., pp. 194–7.

44. Directorate General of Oil and Gas (MIGAS), May 1978.

45. *PE*, November 1978, p. 487.

46. The dates are taken from Bartlett, op. cit., p. 226, but the Indonesian view is presented in Figure 2.2 and was discussed in an interview with Pertamina officials in May 1978. Another branch of the Indonesian government lists the 1961 contract as the first production-sharing contract (see listings by the Directorate General of Oil and Gas). Others may argue, however, that in a strict sense it would be incorrect to consider these contracts to be the prototype for the production-sharing contract as it is now known, because their form was at a rather underdeveloped stage. See, for example, Roderick O'Brien, *South China Sea Oil: Two Problems of Ownership and Development*, Occasional Paper No. 47 (Singapore: Institute of Southeast Asian Studies, 1977), p. 68. O'Brien cites Bartlett's descriptions of the developments of the Asamera and Refican contracts; see Bartlett, op. cit., pp. 226–9.

47. See ibid., p. 226.

48. A history of the evolution of the current government structure in the Indonesian oil industry is given in Bartlett, op. cit., and in Sevinc Carlson, *Indonesian Oil* (Washington, D.C.: Center for Strategic and International Studies, 1976).

49. These points were brought out in an interview with an Indonesian government official in May 1978.

50. Information supplied by Indonesian government.

51 See Section V.1.2. of sample contract, Appendix B.

52. Directorate General of Oil and Natural Gas (MIGAS). See also Table 5.9.

53. This specific information was obtained from Pertamina and Caltex. An example using the formula is in Table 3.2.

54. *Far Eastern Economic Review*, 13 August 1976, p. 13. (Henceforth, this publication will be cited as *FEER*.)

55. *PN*, January 1975, p. 48.

56. See Table 5.1.

57. For an analysis of the Pertamina crisis, see A. Goldstone, 'What was the Pertamina crisis?', in *Southeast Asian Affairs 1977* (Singapore: Institute of Southeast Asian Studies, 1977).

58. Conversation with Indonesian officials, May 1978.

59. This is based on reports in *PN*, August 1976 supplement, and on the discussion in U.S. Embassy, *Indonesia's Petroleum Sector 1977* (Jakarta: July 1977), pp. 15–17.

60. Indonesian government officials, May 1978.

61. Summarized in Rev. Ruling 78–222, 9 May 1978. See Appendix C.

62. See U.S. Embassy, *Indonesian Petroleum Sector*, 1977, p. 19.

63. The dividend tax is a tax on interest, dividends, and royalties. It is imposed at the rate of 20 per cent of taxable profit after deduction of the corporation tax of 45 per cent of taxable profit. See U.S. Internal Revenue Ruling 78–222, released as News Release IR-1991, dated 9 May 1978. (A copy of this document is included as Appendix B of this study.)

64. See Chapter IV, footnote 39, for a computation.

65. In May 1978, Indonesian government officials expected that one of the existing contracts would fall under this category.

66. This information was received from Indonesian government officials in May 1978. It also appears that this had been implemented as early as 1977 in specific cases. For example, the special joint venture between Pertamina and Continental Oil in a block of Irian Jaya was signed by Conoco Irian Jaya Company. See the contract signed 22 October 1977.

67. *AWSJ*, 7 February 1977, pp. 1, 10.

68. *AWSJ*, 21 February 1977, p. 3, and U.S. Embassy, op. cit., p. 17.

69. U.S. Embassy, op. cit., p. 17.

70. This incentive was incorporated in the joint contracts signed by Total and Conoco for Pertamina blocks.

71. *AWSJ*, 20 May 1977, p. 3, and U.S. Embassy, op. cit., p. 18.

72. Indonesian government, May 1978.

73. See, for example, *AWSJ*, 11 March 1977, pp. 1, 5.

74. *AWSJ*, 14 April 1977, p. 3.

75. The following fees were to be paid: (1) inspection fee of data on Block A (Irian Jaya) = \$5,000; (2) inspection fee of data on Block B (Irian Jaya) = \$10,000; (3) fee for tender to participate = \$15,000.

76. *AWSJ*, 6 July 1977, p. 3.

77. *AWSJ*, 19 May 1977, p. 3.

78. Some of the terms of the agreement are the following:
(1) Total, which would operate the Bird's Head block, paid a signature bonus of US$6 million and would spend US$23 million over the first three years. Total would also pay US$2 million when daily production averages 100,000 barrels per day over a 120-day period, and US$3 million for average daily production of 200,000.
(2) Conoco, which would operate the adjoining block, paid a signature bonus of US$3.15 million and would spend US$15 million over the first three years. In addition, the Conoco contract provided for a production bonus to the government of US$1 million for production averaging 50,000 barrels per day over a 120-day period and US$1 million for production averaging 150,000 barrels per day. (This contract information was obtained directly from the Indonesian government. See also *AWSJ*, 24 October 1977, p. 3, and *PN*, January 1978, p. 25; the *PN* report is inaccurate in some of the details on production bonuses.)

79. See Section XV of Conoco's contract and Exhibit 'E' in Total's contract.

80. *AWSJ*, 6 July 1977, p. 3.

81. *ST*, 14 December 1977, p. 6.

IV

The Contractual Framework in Petroleum Exploration and Development in South-East Asia

THE institutional framework within which a firm[1] operates affects its profit calculus, and therefore the firm's estimates of the viability of a petroleum project. This framework includes the laws, policies and regulations that affect the manner in which exploration activities are conducted, the share of the host country in the production, the process of production itself, as well as transportation or movement of the output within or away from the country (including policies on conservation and the environment). The framework also includes policies that affect the inflow and outflow of capital funds. In the past policies affecting a firm interested in exploring for and developing a host country's resources were generally embodied in many separate pieces of legislation, rules, and regulations. It has become the trend, however, to try to embody all such policies in one legislation or specify them in the contracts concluded between the parties.

To appreciate the link between property rights, firm behaviour in South-East Asia, and the resultant pattern of petroleum reserves supply, an explanatory discussion on the contractual framework in the countries of the region would be useful. This chapter presents a short discussion on the general types of contracts, followed by a short outline of each country's contractual terms and conditions. These conditions are further summarized in comparative form in Table 4.1.

No summary will be given for Singapore. Singapore's Petroleum Act covers only the import and export of petroleum and the related

TABLE 4.1
COMPARATIVE SUMMARY OF CONTRACTUAL TERMS IN SOUTH-EAST ASIA, AS OF 1 JUNE 1978[a]

Country/Details	Brunei	Burma	Cambodia (Kampuchea)	Indonesia	Laos	Malaysia	Philippines	Thailand	Vietnam
Contract type	Concession	Concession (onshore) Service contracts (offshore)	Concession	Production-sharing contracts	Concession	Production-sharing contracts	Service contracts	Concession	Service contracts
Area limits	None	Varies according to licence	5 000 km² (onshore) 10 000 km² (offshore)	Variable	—	Variable	50 000–750 000 ha (onshore) 80 000–1 million ha (offshore)	50 000 km² (onshore) Variable (offshore)	NA
Duration	38 years (onshore) 40 years (offshore) (renewable)	± 30 years maximum	Up to 100 years	30 years maximum	Up to 90 years	Maximum 24 years for oil, 34 years for gas	Maximum 50 years	Up to 42 years	>10 years
Relinquishments	Mandatory	Mandatory for SC	Mandatory	Mandatory	Mandatory	Mandatory	Mandatory	Mandatory	NA
Rentals	—		✓	—	✓		—	—	
Work obligations	Minimum expenditures and fixed annual payments	Minimum expenditures	Minimum expenditures	Minimum expenditures	As stipulated	Minimum expenditures	Minimum expenditures and drilling	Minimum expenditures	Minimum expenditures, and activities

Country/Details	Brunei	Burma	Cambodia (Kampuchea)	Indonesia	Laos	Malaysia	Philippines	Thailand	Vietnam
Bonus	—	—	—	Variable	—	Variable	Variable	—	NA
Signature	—	Training of nationals + fee	—	—	—	✓	Equipment & training	—	NA
Discovery	—	✓	—	✓	—	✓	✓	—	NA
Production	—	✓	—	✓	—	✓	✓	—	NA
Royalty	8–12.5%	✓	12.5–15.0%	—	✓	10%	35% ±	—	—
Taxes on profit	50%	—	50%	56%	—	45%	50–70%	50–60%	—
Cost recovery	—	Up to 40%/year from current production	—	Double-declining method, 7–14 years	—	Maximum 20%/yr. on oil, 25%/yr. on gas	—	—	✓ (Including interest-bearing repayment of development costs)
Company's share of 'profit oil'	—	30%	—	15%	—	30%	35–40%	—	'buy-up' privileges up to 45%
Management control	—	State company	—	State company	—	NA	Operator	State	—
Marketing of output	—	—	—	Contractor	—	NA	Optional	Operator	—
Others									
Domestic allocation	—	10% maximum	—	To 25% at 20¢/bbl.	—	—	At world prices	—	—
Pricing	—	—	—	State controlled	—	State controlled	At world prices	State controlled	State controlled
Ownership of equipment	—	—	—	State	—	NA	Operator	—	Operator

[a] Except for Vietnam, which includes developments after June, 1978.

NA = Not available.

Source: Based on descriptions in text of Chapter IV.

aspects of transport and storage.[2] Basic mining laws are carry-overs from laws in force when Singapore was the administrative capital of the Straits Settlements; hence, they do not provide for petroleum exploration and development which are more recent concepts.[3]

General Discussion

An important element of legislation governing exploration and production is the philosophy governing ownership of the resource. In most countries exploration and development rights are granted by the state under some form of contract—a permit, a licence, a joint venture arrangement with the state, a concession, or more recently, a production-sharing contract (which some legal experts see as a mere mutation of the concession).[4] The United States is one of the few countries in which ownership of minerals in the subsoil of privately-held lands remains with the landowner.

In South-East Asia, three basic patterns of work agreements have emerged. These are: (1) the concession, (2) the production-sharing contract, and its variants—the service contract and contract of work, (3) the joint venture agreement with the host government. Each of these recognizes the host nation's ownership of its natural resources as well as the host government's power to control rates of production and pricing of crude oil and its rights to maximize its financial benefits.[5] Each also confines the risks to the private developer.

CONCESSIONS

Concessions include the following features:[6]

(1) Awards are made on the basis of international tenders. Financial considerations include signature and production bonuses,[7] options for local participation (either by the private sector or by the government), contributions to scholarship funds, training of local personnel, etc.

(2) Exploration terms include rentals,[8] minimum work requirements and mandatory relinquishment conditions.

(3) Royalties[9] and income taxes are to be paid.

(4) Drilling and production programmes, safety practices, pricing levels, marketing, etc. are subject to control by the host government.

(5) Provisions are made for contract review under conditions of structural change.

(6) Termination on reasonable grounds by the host government is provided for.

(7) The concession holder normally receives property rights over the surface area of the contract block, and title to the petroleum passes to him at the wellhead.[10] However, in the 'Philippines, the Petroleum Act of 1949 specified that ownership did not pass to concession holders. (See Article 9.)

Two main features of the concessions granted prior to the Second World War were (1) the large areas involved, and (2) the long durations, ranging from 60 to 75 years. Concessions after the Second World War included the concepts of relinquishments of areas to decrease the acreage held and reduction of their duration to about 30 or 40 years.[11]

JOINT VENTURES

In recent years most South-East Asian governments, following the pattern elsewhere in developing countries, have formed state-owned oil companies to undertake exploration for and development of petroleum resources—either singly or jointly with private companies.[12] Pressures from host governments to enter into partnerships with private companies entail minor adjustments, as joint venture arrangements between petroleum companies, a practice in existence for some time, have become even more common in recent years. Even production-sharing contracts may involve joint venture arrangements. The increased frequency of such arrangements have been motivated by several factors. Among these are: the need among relatively smaller firms to harness substantial amounts of capital resources, to spread the risks of exploration among both small and large firms, to unitize operations in a common field, to take advantage of economies of scale in certain activities, to avoid unnecessary investment costs in downstream operations, or to provide its downstream operations with an assured supply of crude.

Joint ventures basically cover the following areas:
  (1) Sharing of costs and profits in exploration and development, including minimum work obligations.
  (2) Control over such operations.
  (3) Legal structure of operations following commercial discovery, including the obligations of the partners.
  (4) Taxation and royalty rates.
Joint venture arrangements usually run for around 25 years.

SERVICE CONTRACTS, PRODUCTION-SHARING CONTRACTS, AND CONTRACTS OF WORK[13]

In this type of arrangements, ownership of the petroleum resources by the state is explicitly recognized by the signatories; exploration and development of such resources are conducted under contracts negotiated with a host government agency. The contractor provides the risk capital, equipment, and technicians to carry out the exploration and development programme without compensation in case of failure to develop commercial production.

Indonesia has the distinction of having pioneered the production-sharing contract. The difference between a service contract and a production-sharing contract is subtle. It is sometimes defined as the manner in which the contractor acquires a portion of the crude output.[14] This distinction is sometimes not clear; see, for example, the Philippines service contract in this chapter. In the service contract, the contractor acquires a portion of the output at an agreed price, perhaps below the prevailing market or 'realized' price, after obtaining part of the annual production to recover some or all of its costs. Under the 'production-sharing contract', the operator receives a portion of the crude to recover its exploration and operating costs, also referred to as 'cost oil', plus a portion of the balance, which is referred to as 'profit oil'. 'Profit oil' is subject to existing taxes on income. In Indonesia, in addition to the above split of the crude oil output, the contract requires that a portion of annual gross output be sold to the national oil company at a stipulated preferred price, i.e., at prices below prevailing market levels.

With the exception of the above distinctions, both types of contracts have common features that include the following:

(1) The contractor provides all required financing, technical expertise and equipment to carry out the exploration and development programme. (In the Indonesian case, ownership of the equipment is transferred to the government; compensation is recovered from production, if any.)

(2) There are 'minimum expenditure' requirements, relinquishment provisions, and payments of bonuses and other types of financial obligations.

(3) The contractor markets any exported portion of the host government's share of production.

(4) There are additional requirements on employment and retraining of nationals, as well as provisions on management control. The final say in the conduct of operations on work programmes may be embodied in the host country's representative, e.g., the national company.

(5) In the Indonesian case, the title to the petroleum passes to the contractor at the point of export.

The Indonesian 'contract of work' (*kontrak karya* or, more correctly—according to the legalists in Indonesia—the *perjanjian karya*) is sometimes categorized as falling under the general heading of 'service contracts'.[15] These contracts replaced the concessions of Caltex, Shell, and Stanvac in 1963, as the concession system became extinct under Indonesia's Law No. 44 (see the section on Indonesia below, pp. 86–90). These contracts, however, gave the contractor complete control over its operations, in terms of management and marketing, and because of this the *kontrak karyas* have sometimes been alluded to as being 'disguised concessions'.[16]

Brunei[17]

The basic legislation now covering petroleum operations in Brunei are 'The Petroleum Mining Enactment, 1963' and 'The Income Tax (Petroleum) Enactment, 1963'. These were amended by 'The Petroleum Mining (Amendment) Enactment, 1969' and 'The Petroleum Legislation (Amendment) Enactment, 1972'. Definitions on ownership of offshore resources in the 1963 mining legislation are based on Proclamation No. S.41, 1954, on the continental shelf.

Annexed to the Petroleum Mining Enactment, 1963, are model petroleum agreements covering both onshore and offshore types of concessions. These models spell out the conditions under which such concessions may be held.

1. *Area limit*: none.
2. *Duration*: 38 years onshore, 40 years offshore, with renewals for a 30-year period.
3. *Relinquishments*: 50 per cent after 8 years onshore, and after 10 years offshore; additional 25 per cent (or total 75 per cent) of the original area after 15 years onshore, and after 17 years offshore.
4. *Work expenditure obligations*: Minimum expenditures of specified amounts required according to acreage held during the first 8 years over two 4-year periods for onshore areas, and over two 5-year periods during the first 10 years for offshore areas. Fixed annual payments are required for subsequent years, such payments being deductible from royalties. (See Part VI of model contracts in the enactments.)
5. *Royalties*: 12½ per cent of value or volume of output onshore, i.e., in areas within territorial waters of 3 miles' width; and 10 per cent and 8 per cent in areas beyond territorial waters onto the edge of the continental shelf, depending on the distance from the shoreline.
6. *Tax*: 50 per cent of net profit, i.e., total proceeds minus costs, including annual fixed payments. Royalties are considered part of the 50 per cent tax.

In addition, companies obtain *exclusive rights* over their operations and over discoveries. Subject to certain limitations, companies may use other resources required by their operations (e.g., enter adjacent lands, erect offices and buildings, cut timber, use water, etc.). (See Part III of the model contracts in the legislation.)

Burma

No information is available at this writing on the existence of any legislation applicable to contracts covering exploration and development. With the enactment of the Enterprise Nationalization Law in 1963, the state nationalized all private operations and assigned their assets to the state-owned company, Myanma Oil Corporation. The *Petroleum News* (January 1979) reports that concessions in Burma are still subject to the Burma Petroleum Concession Rules of 1962, under the terms of the Petroleum Resources Act of 1957. In 1974, however, offshore blocks were offered and leased out by the government of Burma to foreign contractors for exploration and prospective development. These blocks were operated by these foreign contractors on service contracts patterned after the production-sharing contract of Indonesia. In December 1978, the government again called for exploration bids from international companies 'for newly-drawn up offshore blocks. The contracts covering these new blocks were similar to those offered in 1974.[18] Concessions—if granted—would then most likely apply only to onshore areas. At this point, this would be academic inasmuch as onshore operations have been strictly state-run.

Burma's Petroleum Concession Rules, 1962, are subject to the stipulations in the Petroleum Resources (Development and Regulation) Act of 1957.[19] The main features of the various petroleum licences and the service contract for offshore areas are outlined below:

PETROLEUM EXPLORING LICENCE
1. *Area limit*: 810 sq. km.
2. *Duration*: 2 years, renewable for the third year for 50 per cent of the original area, and a fourth year—at the discretion of the Financial Commissioner.
3. *Rentals*: set per square mile per year at $1.94 for the first 2 years, $3.88 for the third year, and $5.83 for the fourth year.
4. *Work expenditures*: for the first year: $0.10 per acre or $4,854 whichever is greater, rising to $0.15 per acre or $9,708.

PETROLEUM PROSPECTING LICENCE
1. *Area limit*: 100 sq. km (?).[20]
2. *Duration*: 3 years, renewable for a fourth year, and a fifth year—at the discretion of the Financial Commissioner.

3. *Rentals*: set per acre per year at $0.19 for the first year, rising to $0.49 in the fourth year.
4. *Work expenditure*: $58,000 rising to $97,087.
5. *Conversion to mining lease*: limited to 60 per cent of prospecting area.

PETROLEUM MINING LEASE[21]

1. *Area limit*: 1 300 sq. km (?) (not to exceed 60 per cent of prospecting area?).
2. *Duration*: 20 years, extendable for another 20 years.
3. *Rental*: set at $0.97 for the first 130 sq. km, $1.94 for the second 130 sq. km, $3.88 for the next 259 sq. km and $5.83 for the next 647 sq. km.

THE SERVICE CONTRACT[22]

1. *Area limit*: The limit for production is 25 per cent of the original contract area.[23] An additional 25 per cent is retainable for further exploration and future production.
2. *Duration*: 3 years exploration period, with possible extension of another 3 years; a development period following commercial discovery; 20 years production period following the development period.
3. *Relinquishment*: 75 per cent of original area by the time development of a commercial discovery begins.
4. *Work expenditure obligations*: stipulated minimum requirement over initial defined period (e.g., Cities Service—$7.5 million per block initial 3 years; Esso—around $18 million per block initial 3 years). A specified number of wells may also be required although the model contract does not specify so.
5. *Bonuses*: a) No signature bonus required, but training of nationals is required upon signature; also $1 million (negotiable) fee for the purchase of data.

   b) Discovery[24] and production bonuses are levied on a sliding scale from $1 to $10 million.

   c) Data are acquired by purchase from Myanma Oil Corporation, such data costing as much as $2 million per block. In the 1978 invitations to bid, the government would not release data on the blocks even with cash payment, such data to be released only with payment by the *successful* bidding company.[25]

6. *Cost recovery*: up to 40 per cent of production per year for exploration, development and operating costs.

7. *Profit split*: 70/30 in favour of Myanma.[26]

8. *Taxes*: The government's share of production included taxes due the state from the contractor.

9. *Domestic allocation*: maximum 10 per cent.

10. *Management*: reserved to the state company.

11. *Others*: In addition, the contractor provides all financial requirements, expertise and technology, and shoulders all risks of exploration (with no compensation or reimbursement of expenditures). All equipment become the property of Myanma Oil Corporation. The state company also holds title to all data acquired from the contractor's operation.

Cambodia (Kampuchea)[27]

Prior to the end of the Indochina War, Cambodia's basic legislation governing the development of petroleum resources was Mining Law 380/68-CE of 16 December 1968, which covers exploration and development of all natural deposits on land on Cambodia's continental shelf. Other related laws were: No. 221-NS of 13 September 1957, No. 242-CE of 18 June 1964, and No. 463-71-CE of 2 March 1971. Law No. 463-71-CE and its accompanying decree No. 199/71 of 13 April 1971 concern the fiscal aspects of petroleum operations. No information is available at this writing on existing petroleum policy, but the provisions of the old legislation are listed below:

1. *Area limits*: 5 000 sq. km onshore and 10 000 sq. km offshore.

2. *Duration*: a) Exploration permit (referred to as Permit H): 5 years, with extension allowed if work expenditures meet requirements and if further similar expenditures are agreed upon by the contractor. The permit is converted to a concession if discovery is made.

b) Concession (exploitation permit, also known as Concession H): 40 years, renewable for two 25-year period extensions. A concession may be terminated prior to the expiration date.

3. *Relinquishment*: 50 per cent at the end of 18 months and additional 25 per cent between fifth and eighth years of the exploration period; 50 per cent when renewing the permit.

4. *Exploration work expenditures*: first 5-year period of exploration—$1 million minimum on each block.

5. *Rentals*: from $30/sq. km for the first 5 years to $400/sq. km during the last quarter of the 40-year concession.

6. *Royalties*: 12.5 per cent of production below 5 million tons/year and 15 per cent for production greater than 5 million tons/year.

7. *Tax*: annual net profit tax. If the production royalty is less than the taxable profit, then the difference is subject to a direct tax of 50 per cent or more. Losses may be credited against gross profits and carried over for 3 years over the first 10-year period.

The Cambodian government reserved the right to acquire a 20 per cent working interest in a concession, reimbursing the company for proportional expenses out of production.

Indonesia

Indonesia's present basic legislation governing petroleum exploration and development and all related activities is Law No. 44 of 1960. This law contains the philosophical basis for the contractual arrangements, i.e., state ownership of petroleum resources (Article 2) and state mining of such resources, either directly or through contractors (Articles 3 and 6). Some of the more important pertinent legislation are: Law No. 11, 1967, on the basic provisions of mining; Government Regulation No. 32, 1969, on the implementation of Law No. 11, 1967; and Law No. 8, 1971, on the State Oil and Natural Gas Mining Enterprise (including its amendment in Law No. 10, 1974). Also pertinent are the 'Government Regulation re Indonesian Territorial Waters' (Law No. 4 of 1960) and the 'Government Proclamation over the Indonesian Continental

Shelf', 17 February 1969 (Law No. 1 of 1973).[28]

Law No. 8 of 1971 centralized the responsibility over oil and natural gas mining activities in the Department of Mining and established Pertamina (short for 'Perusahaan Pertambangan Minyak dan Gas Bumi Negara' which translates to State Oil and Natural Gas Mining Enterprise). Pertamina was the legal entity entitled to carry out oil and natural gas mining. (See Law No. 8, Chapter I, Article 1, and Chapter II, Article 2.)

Except for areas held under concessions at the time the concession system was abolished, non-state operations are now conducted under what are called 'production-sharing contracts'. In the meantime, the former concessions were replaced by what are referred to as 'contracts of work'. Production-sharing contracts on Pertamina-held blocks were introduced in 1977.

Summarized below are the main elements of the different contracts that have been in operation in Indonesia over the last decade or so.

CONTRACTS OF WORK[29]

1. *Duration*: 30 years for new areas broken down as follows:[30]

   10 years exploration;

   20 years exploitation.

   Contracts for older areas run for 20 years and cover only production or exploitation.[31]

2. *Relinquishments*: 25 per cent after first 5 years of exploration; total 50 per cent by the end of 10 years.

3. *Work expenditures*: minimum investments specified.

4. *Bonuses*: signature, $5 million for each area; production, $5 million for each area.

5. *Profit split*: 60/40 in favour of the state. The state could elect to take 20 per cent of aggregate production in kind.

6. *Domestic allocation of crude*: Computed 'off the top' or from gross output.[32]

7. *Marketing*: by contractor.

8. *Management control*: by contractor.

Other provisions concerning take-over of property and training of personnel were included.

PRE-1976 VERSION OF THE PRODUCTION-SHARING CONTRACT

1. *Duration*: 30 years, broken down as follows:

   10 years exploration;

   20 years exploitation

   a) Contract would automatically be terminated with non-discovery after 10 years, but exploitation period would commence upon discovery.

   b) Option to terminate earlier was negotiable, but termination could not take place earlier than 2–3 years.

2. *Relinquishments*: mandatory or voluntary, but negotiated beforehand.

3. *Work expenditures*: negotiated, with specified minimum periods and amounts.

4. *Bonuses*:[33] a) Signature bonus, and compensation for data on acreage.

   b) Production bonus, after reaching certain production levels during a period of 90–180 consecutive days.

5. *Taxes*: Income taxes were paid, according to existing laws and regulations. The state enterprise's share of crude included payment of the contractor's income tax obligations, which the state enterprise paid to the Indonesian government. Dividend taxes were also required and included in the government share.

6. *Cost recovery*: Costs were recovered up to 40 per cent of annual production; the balance of costs were deferred for future production.

7. *Profit split*: a) The total 'profit oil' was split initially at 65/35 in favour of the state enterprise, and at 67½/32½ upon reaching certain production levels (50,000 b/d, 75,000 b/d, or 100,000 b/d, depending on field size).

   b) *Domestic allocation*: The operator would provide the domestic market with a certain percentage of output prorated on the basis of industry production; this would amount to not more than 25 per cent of its 'profit oil' share (35 or 32½ per cent, as the case may be), and the company would sell this portion to the government at a price of 20 U.S. cents per barrel at the point of export.[34]

8. *Management control; including approval of the work programme and budget*: the state enterprise.

9. *Marketing of output*: to be done by operator, including marketing of the state's share with certain exceptions.

10. *Others*: a) Title to the contractor's portion of crude output was received at the point of export.

   b) The operator provided all the necessary funds, equipment, foreign exchange, technical aid, etc. The equipment became the property of the state enterprise upon entering Indonesia, and compensation would be recovered, along with other costs, from production.

REVISED PRODUCTION-SHARING CONTRACT, 1976[35]

The following changes to the earlier production-sharing format were introduced:

1. *Production bonus period*: 120 days (compared to 90–180 day variable in earlier contract).[36]

2. *Taxes*: according to existing laws and regulation. This remained at a total of 56 per cent, composed of 45 per cent corporate tax and 11 per cent dividend tax.[37]

3. *Cost recovery*: a) The double-declining depreciation method is used for capital expenditures. Recovery is on a 7-year basis for Group I companies, and on a 14-year basis for Group II companies.[38]

   b) The straight-line method, with 8 per cent interest, is used on non-capital expenditures recovery, on the same period bases as above.

4. *Profit split*: The pre-tax split of the 'profit oil' is 65.91/34.09 in favour of the state enterprise. The state enterprise's share does not include taxes or domestic allocation portions due from the contractor. When taxes are included, the effective profit split is 85/15.[39] When allocation for the domestic market is further deducted, the final net 'take-home' crude is less (about 12 per cent);[40] since this portion is sold to the government at only 20 U.S. cents per barrel f.o.b. at the point of

export, an income that is negligible compared to earnings from the rest of the crude, it may not properly be counted in with the rest of the company's 'profit oil'.

JOINT VENTURE ARRANGEMENT[41]

Joint venture arrangements between the state enterprise and private enterprises were introduced in 1977. Their main features were as follows:

1. *Location*: areas reserved for the state enterprise.
2. *Payments*: a) *Compensation payment* to inspect and acquire data.

    b) *Signature bonus* to participate in the venture.
3. *Profit structure*: a) *Costs*: The state enterprise and the private partner share the costs of exploration, development and production on a 50–50 basis. The partner matches the state enterprise's previous expenditures, or bears the balance of the costs of exploration until the end of 3 consecutive years of operation, whichever comes later.

    b) *Profit split*: 50–50 split of oil or gas output followed by the standard production-sharing contract terms to cover the 50 per cent share of the partner.

Laos

Very little information is available on Laos. The best information available is dated 1971 and was presented at a UN seminar in Bangkok.[42] According to the proceedings, petroleum exploration and development were covered in Laos by general mining legislation. These were basically the Royal Ordinances No. 42, 26 January 1959, and No. 161, 26 May 1959, plus supplementary decrees. Earlier statutes were also in force; these included Law No. 57–26 of 30 September 1957 (which provided for fiscal aspects), amended by Law No. 65/15 of 27 October 1965.

The basic contracts in the upstream stages of the petroleum industry under these legislation were the *exploration permit* and the *concession* which covers the exploitation stage. Their basic features were as follows:

1. *Duration*: a) *Exploration*: 5 years initially, renewable for

another 5 years. Discovery entitles holder to a concession.

   b) *Exploitation concession*: 40 years, renewable.

2. *Relinquishments*: 50 per cent of area at the beginning of each renewal at the exploration stage.

3. *Work expenditure obligations*: stipulated annual work obligations at the exploration stage; details not available.

4. *Rentals*: annual surface rentals required; amounts unknown.

5. *Royalty*: required but not known at this writing.

Malaysia

Petroleum exploration, development, and marketing activities in Malaysia are currently governed by the Petroleum Development Act, 1974, and the Petroleum Regulations, 1974.

The Act vests in the national enterprise the entire ownership in and control over exploration, exploitation, and extraction of petroleum (Act, Section 2) as well as of processing, refining and manufacturing of petro-chemical products from petroleum (Section 6). A private enterprise engaged in any of the above activities at the time of enactment of the Act could continue to do so with permission of the Prime Minister under some type of arrangement (Section 9).

Under the Petroleum Regulations of 1974, all companies with exploration licences or with agreements under the Petroleum Mining Act of 1966 were to make available to the state enterprise, without compensation, *all* data obtained on Malaysian acreage acquired prior and subsequent to enactment of the 1974 Act.

The *production-sharing contracts* signed with Exxon and Shell in 1976 have the following features:[43]

1. *Duration*: 20 years, with possible 4-year extension for oil and 14-year extension for gas. The 20-year period is divided as follows: 2 years exploration, 3 years development, 15 years production.[44]

2. *Royalties*: total 10 per cent, broken down into 5 per cent of gross production of oil or of gross sale of gas to the Federal government, and 5 per cent of gross production of oil or gross sale of gas to the appropriate state government. (Royalties

are deducted from output prior to splitting the 'profit oil'.)

3. *Bonuses*: signature bonus, information bonus for data provided with signature, discovery bonus of M$2.5 million (about US$1 million) for a commercially viable field, production bonus of M$5 million (about US$2 million) when output reaches 725 000 kilolitres (about 4,560 barrels)[45] per quarter.

4. *Income tax*: 45 per cent on 'profit oil' to be paid by both the contractor and national enterprise to the Federal government.

5. *Cost recovery*: up to maximum of 20 per cent per annum for oil and up to a maximum of 25 per cent per annum for non-associated gas.

6. *Profit split*: the balance of 70 per cent (for oil) or 65 per cent (for gas) is split 70/30 in favour of the national enterprise.

7. *Pricing*: proceeds in excess of US$80/kilolitre (about $12.72/bbl.) would be split 70/30 in favour of the national enterprise.

8. *Research fund*: 0.5 per cent of proceeds from gross proceeds (i.e., from the sale of cost and profit oil).[46]

RELATED LEGISLATION

Under the Continental Shelf Act of 1966, Malaysia proclaimed ownership of the natural resources in submarine areas beyond its territorial waters up to water depths of 200 metres or at greater depths if the water depths allow exploitation of such natural resources. Other related legislation were the Petroleum Mining Act of 1966, the Petroleum (Income Tax) Act, 1967, as amended the same year; and the Petroleum Mining Rules, 1968.

The Petroleum Mining Act and the Petroleum Income Tax Act outlined a conventional concession-type relationship of private companies with the state. Although mostly superseded by the 1974 Act, the main features of the 1966 Act are outlined below for historical perspective:[47]

1. *Duration*: 10 years exploration, with possible 5-year extension. 30-year development period, extendable by agreement.

2. *Relinquishment*: 50 per cent of original area after 5 years; additional 25 per cent, or total 75 per cent, after the next 5 years, unless the remaining 25 per cent was too small to develop.

3. *Work obligations*: Varying range, rises to US$1 million/1 000

sq. m after the tenth year of exploration.

4. *Rentals*: Begin with fifth year, rising in rate in the tenth year of the exploration period, and rising further from the first year of the development period.

Philippines

Petroleum exploration and development activities in the Philippines are now conducted within the framework of Presidential Decree No. 87, which took effect on 21 December 1972, and subsequent decrees. P.D. 87, also known as 'The Oil Exploration and Development Act of 1972', supplemented the Petroleum Act of 1949 in its operational aspects, but not in the basic philosophy governing ownership of resources.

The Petroleum Act of 1949 proclaimed

...ownership by the state of all natural deposits or occurrences of petroleum or natural gas in public and/or private lands in the Philippines, whether found in, on, or under the surface of dry lands, creeks, rivers, lakes, or other submerged lands within the territorial waters or on the continental shelf, or its analogue in an archipelago, seaward from the shores of the Philippines... (Article 3).

P.D. 87 did not alter the above philosophy but expressed as state policy the acceleration of the discovery and production of indigenous petroleum through the utilization of government and/or private resources, local and foreign 'under... arrangements... which are calculated to yield the maximum benefit to the Filipino people and the revenues to the Philippine government for use in furtherance of national economic development...' (Section 2).

The contractual instrument for conducting exploration and development under the Petroleum Act of 1949 was the concession. P.D. 87 introduced the concept of service contracts to cover exploration and development for the government (Sections 4 ff.). With Presidential Decree No. 782, dated 25 August 1975, all concession holders were required to convert their holdings to service contracts by 25 August 1976. Presidential Decree No. 1206, of 6 October 1977, which created the Department of Energy, also introduced certain changes: (1) it forbade the granting of a depletion allowance in service contracts, (2) subject to certain specified exceptions, it defined the minimum and maximum sizes of contract areas, and (3) it defined the minimum annual net revenue or share of the government resulting from the operation of the service contract (Section 12(2)).

THE SERVICE CONTRACT

The Philippine service contract is similar to the Indonesian production-sharing contract. P.D. 87 clearly states that all petroleum produced belongs to the government (Section 6). Payment of the service fee of the contractor and of operating expenses would be made by the government, from funds proceeding from the sale of such petroleum if commercial production is found (Section 7). A significant distinction is that the official party to the contract is a government regulatory agency rather than the state oil company. Originally this agency was the Petroleum Board, subsequently replaced by the Energy Board. The latter was then absorbed in a newly created Department of Energy in 1977, which subsequently was elevated to a Ministry in 1978. Other distinctions (e.g., concerning management, marketing, ownership of equipment, and pricing of the domestic allocation) will be noted below in outlining the provisions of the contract.

1. *Area limit*: onshore, not less than 50 000 hectares nor more than 750 000 hectares; offshore, not less than 80 000 hectares, nor more than 1 million hectares (P.D. 1206, Section 12(2)).[48]

2. *Duration*: 7-year initial exploration period, extendable to 10 years. The contract automatically expires with non-discovery. Two 1-year extensions for exploration are allowed with discovery. With production, a 25-year extension beyond the ·10-year exploration period is allowed, plus additional extensions not exceeding 15 years (P.D. 87, Section 9).

3. *Relinquishments*: 25 per cent of the total area after the first 5 years, and in the case of an extension, an additional 25 per cent at the end of the seventh year. In the producing stage, the contractor may retain the producing area plus 12½ per cent of the original acreage for exploration purposes. He

pays rentals on acreage held (P.D. 87, Section 9(c)).

4. *Work expenditure obligations*:

   a) Minimum expenditures per hectare per year: onshore, at US$3 from the first to the fifth year, and US$9 from the sixth to the tenth year; offshore, at US$3 during the first and second years, US$6 from the third to the fifth year, and US$18 from the sixth to the tenth year.

   b) Drilling commitment (a specific number of wells per year) (P.D. 87, Section 9).

5. *Bonuses*: Signature, discovery and production bonuses (originally absent) were incorporated in later contracts.[49] The signature bonus often takes the form of equipment (such as laboratories) or scholarships to train personnel abroad.[50] The production bonuses follow a negotiable sliding scale, and are imposed at *any* of several stages or production levels over a 60-day period[51]:

   —declaration of commercial discovery
   —on 25,000 b/d production
   —on 50,000 b/d production
   —on 100,000 b/d production
   —on 150,000 b/d production.

6. *Income tax*: The contractor is liable only for income taxes but is exempt from all others. Deductions from gross income for computing taxable net income are: (1) allowance for Filipino participation (which reportedly was removed in later contracts) and (2) operating expenses (Section 21, of P.D. 87).

7. *Cost recovery allowance*: 70 per cent of annual production, with balance to be carried forward.[52] Contracts from 1974 onwards allowed 55 to 70 per cent recovery.[53] Costs incurred in unsuccessful operations are not reimbursable (Section 8(1), P.D. 87).

8. *Government 'take'*: The state should receive no less than 60 per cent of 'profit oil' (gross output minus expenses and Filipino participation allowance). Such 60 per cent share would include all taxes paid by the contractor.[54] Contracts from 1972 to 1974 provided for a 60/40 split. A contract

signed in late 1974 provided for a 62.5 per cent government share, with cost recovery reduced to 55 per cent. Another contract in 1975 provided for a 65/35 split in favour of the state, with cost recovery at 55 per cent of production per year.[55]

9. *Management control*: The operator is expected to 'manage and execute petroleum operations', and the government merely oversees such management (P.D. 87, Section 8). (Note the difference from the Indonesian contract.)

10. *Marketing*: The operator may be authorized to market the output for the government domestically or internationally (Section 8, P.D. 87). The 'posted price' at which petroleum would be offered for sale would be decided by the contractor in consultation with the Petroleum Board (P.D. 87, Section 3).

11. *Ownership of equipment, materials, plants and other installations*: Items of a movable nature remain the property of the contractor unless they are not removed within one year after the termination of the contract (Section XI of model agreement). (Note the difference from the Indonesian contract.)

12. *Other obligations of the contractor*:

   a) The contractor provides all necessary services, technology and requisite financing, assumes all risks for exploration without reimbursement for failure to discover a commercial reservoir, and furnishes all technical information required (Section 1.3 and Section 6.1 of model agreement).

   b) The contractor also provides the domestic market with a portion of the supply at *world* market prices (Section 6.1(m) of model agreement).

   c) Contracts after 11 June 1978 are to include a clause allowing the President of the Philippines to revoke or modify such contract (P.D. 1585).[56]

THE CONCESSION

To provide some historical perspective, the major features of the *concession system* under the old Petroleum Act of 1949 are briefly summarized below:

1. *Area limits*: a) Exploration, not more than 500 000 hectares of exploration area in any one petroleum region, nor more than a total of one million hectares in the whole country.

   b) Exploitation, no more than 250 000 hectares in any region or more than a total of 500 000 hectares in the Philippines (Article 44).

2. *Duration*: a) Exploration: 4 years, with two possible extensions of 3 years each, or a total of 10 years.

   b) Exploitation: 25 years maximum for the first term, renewable for another 25 years (Article 72).

3. *Relinquishments*: Relinquishments are made at the option of the concessionaire, but the total retained portion of one or more contiguous blocks should not be less than 20 000 hectares (Articles 50 and 73).

4. *Work expenditures*: required on both exploration and exploitation concessions.

5. *Royalties*: 12½ per cent (Article 65).

Thailand

The basic legislation governing petroleum exploration and production in Thailand are currently the Petroleum Act of 1971, as amended in 1973 (Petroleum Act B.E. 2514 and Petroleum Act No. 2, B.E. 2516), and the Petroleum Income Tax Act of 1971.[57] The areas covered by the petroleum legislation were defined as extending over Thailand's continental shelves, as demarcated by proclamation on 18 May 1973.[58] The 1973 amendments to the 1971 Act covered exceptions applicable to offshore blocks in waters of 200-metre depth. Such blocks received special considerations in area size, relinquishment requirements, royalty rates, etc., in recognition of the higher risks and larger capital investment requirements involved in deep-water exploration and production.[59]

The legal contract for exploration and development in Thailand is a concession, and the explicit form is as prescribed in various ministerial regulations.[60] The two sets of legislation provide for the following:

1. *Area limits* (PA, Section 28): a) Four to five blocks, whose

aggregate area must not exceed 50 000 sq. km, except as in (b).

b) In water depths greater than 200 metres: size is not restricted to 50 000 sq. km but may be enlarged by the Minister.

2. *Duration*: a) The *exploration period* is 8 years or less, with one renewal period of up to 4 years, provided the initial exploration period was specifically not to exceed 5 years. The exploration concession does not expire with commercial production (PA, Sections 25, 26, and 41).

b) The *production period* may not exceed 30 years, following the termination date of the exploration period, notwithstanding any production in the exploration period. The production period may be extended once for a period not exceeding 10 years.

c) The concession expires with non-discovery at the end of the renewed exploration period (PA, Section 36).

3. *Relinquishments*: a) Except as provided for in (b) and (d), 50 per cent of a block must be relinquished at the end of the fifth year, and a further 25 per cent if the exploration period is extended.

b) In offshore blocks with waters deeper than 200 metres, except as in (d), 35 per cent of the block must be relinquished at the end of the fifth year, and a further 40 per cent if the exploration period is extended (see PA, Section 36, as amended).

c) Total relinquishment is required at the end of the second petroleum exploration period (see PA, Section 36).

d) The area in any block retained for commercial production would be the production area plus 12.5 per cent of the initial block for further exploration (PA, Section 45).

4. *Work obligations and expenditure requirements*: These are divided into three obligation periods during the exploration period; the first covers the first 3 years of the exploration period, and the third covers the extension period (PA, Section 31).

5. *Royalties*: The Petroleum Act, including its amendments, provides for the following:

a) *Cash payment*: 8.75 per cent of value sold or disposed of in offshore blocks in water depths greater than 200 metres; 12.5 per cent, otherwise.

b) *In volume*: volume equivalent to 7/73 (or about 9.6 per cent) of the value of petroleum sold or distributed, for offshore blocks in water depths greater than 200 metres; volume equivalent to one-seventh of the value, otherwise. (See PA, Section 84, as amended.) Royalties are expensed (see PITA, Section 24).

6. *Taxes*: The concessionaire is exempt from payment of all kinds of taxes, duties, fees, and levies except income tax, royalties for timber or petroleum, and for fees for services rendered as provided for in the Act (PA, Section 71). In addition, 'special advantages' voluntarily made by the concessionaire prior to the award are taken into account in the allocation of blocks among applicants.[61] The net profit tax is not less than 50 per cent and not more than 60 per cent, with losses carried for up to 10 years (PITA, Section 20).

7. *Other benefits and rights of the concessionaire*: a) The law provides that the state shall not nationalize the concessionaire's properties and his rights to conduct petroleum exploration (PA, Section 64).

b) The concessionaire may own land necessary for his operations (PA, Section 65) and may use—under certain conditions—land or property adjacent to his concession that may be required for his operations (PA, Sections 66, 67, 68).

Vietnam

No information is available on any specific legislation or body of legislation covering petroleum exploration/exploitation in reunified Vietnam. A decree, No. 115/CP[62] enacted 18 April 1977, outlines the general policy on foreign investment.[63]

The new service arrangements entered into in 1978 between the Vietnamese governments and three exploration groups—Deminex, Agip (ENI), and Bow Valley[64]—are reported to provide for cost recovery from any oil discoveries and 'buy-up' privileges of up to

50 per cent of production at 5–11 per cent below the prevailing world price.[65] Exploration costs were reported to be reimbursable from any oil discovered, while investments in development would be repaid with interest at market rates 10 years after the start-up of production.[66]

A contract signed with a Norwegian company reportedly provides for recovery of the capital funds, costs of equipment, and cost of services from oil production, if successful, or by repayment by the government, otherwise.[67]

The contracts cover offshore areas which appear to be in the size range of 15 000 sq. km. AGIP's contract area was reported to total 15 000 sq. km, while Bow Valley's acreage covers around 14 000 sq. km.[68]

Earlier arrangements were reported between North Vietnam's government and ENI, the Italian state company. An agreement signed on 18 April 1973 provided for joint prospecting along North Vietnam's coastline and continental shelf, advanced technical training for Vietnamese nationals, the erection of downstream facilities as well as trade agreements in non-petroleum sectors.[69] Other unconfirmed reports circulated concerning agreements with Japanese companies. Very little substantial information is, however, available on contracts in that part of Vietnam either before or after reunification, and none on legislation.

Petroleum exploration and exploitation in South Vietnam prior to reunification of the north and south was governed by Law No. 011/70 dated 1 December 1970, which set out the general framework for exploration and exploitation concessions. There are other related decrees and orders: Decree No. 003-SL/KT of 7 January 1971 established a National Petroleum Board. Ministerial Order No. 249-BKT/VP/UBQGDH/ND of 9 June 1971 delimits the Republic's continental shelf claim and provides the procedures to be followed by companies interested in offshore exploration and exploitation. Decrees No. 062-SL/KT of 2 April 1973, No. 064-SL/KT of 2 April 1973, No. 065-SL/KT of 2 April 1973, No. 170-BKT/VP/UBQGDH/ND, No. 171-BKT/VP/UBQGDH/BD and No. 172-BKT/VP/UBQGDH/ND, all dated 17 April 1973, define the conditions pertinent to the concession contract. For historical purposes, and to provide points of reference, the conditions under the old concession are summarized below:[70]

1. *Area limit*: a) Exploration: 20 000 sq. km maximum per concession, and not more than five blocks per concessionaire.

   b) Exploitation: 500 sq. km maximum, and maximum ten blocks per concessionaire.

2. *Duration*: a) Exploration: 5 years initially, renewed automatically for another 5 years, if conditions have been fulfilled, and extendable for a third 5-year period at the discretion of the government.

   b) Exploitation: 30 years, extendable for an additional 10-year period.

3. *Relinquishments*: a) Exploration: Originally 50 per cent had to be relinquished at the end of the first 5 years, and a further 25 per cent at the end of the first extension period.[71] This was reportedly changed by Decree No. 172-BKT/VP/UBQGDH/BD dated 17 April 1973.[72] This decree required relinquishment not later than the end of the third year, and a minimum relinquishment of 500 sq. km per sub-area. Total relinquishment was required at the end of the fifteenth year should the block not be converted to an exploitation concession.

   b) Exploitation: 50 per cent at the end of the first 5 years with no production; total relinquishment at the end of ten years from the original granting date with no production; total relinquishment after 2 years from cessation of production for reasons other than *force majeure*.

4. *Rentals*: surface taxes for each type of concession, ranging from US$4 per sq. km or fraction thereof during the first three years of the exploration period, and rising to US$2,000 from the sixth year of the exploitation concession period.[73]

5. *Work obligations and expenditures*: expenditure requirements for each 5-year exploration period, and drilling within one year after an exploitation concession is granted.

6. *Bonuses*: signature, discovery and production bonuses. Production bonuses are payable for production of 75,000 b/d, 100,000 b/d, 150,000 b/d, and 200,000 b/d sustained over 60-day periods.[74]

7. *Royalties*: 12.5 per cent, in cash or in kind or both. Surface tax may be credited against royalties.

8. *Tax*: income tax at a rate varying from 45 per cent to 55 per cent, depending on the desirability and risk of the concession area. Royalties and surfaces taxes are expensed.

1. The term 'firm' is used here to refer to a unit engaged in production for profit, and assumes that this unit's objective is to maximize such profit.

2. See Chapter 256, Volume VII, *The Statutes of the Republic of Singapore*, issued by the Law Revision Commission, Singapore (Singapore: Government Printing Office, 1970), and the Amendment of 2 March 1978, *Government Gazette, Acts Supplement*, 10 March 1978.

3. UN, ESCAP, *Proceedings of the Seminar on Petroleum Legislation with Particular Reference to Offshore Operations*, 1971, Mineral Resources Development Series No. 40, 1973 (henceforth to be referred to as 'MRDS No. 40'), p. 136. (Other publications in this series will be similarly referred to, by MRDS number.)

4. See Robert Fabrikant, for example, in *PN*, July 1973, p. 47, 'Legal Aspects of Production-Sharing Contracts in the Indonesian Petroleum Industry'. Fabrikant wrote:

'At first glance, it is difficult to disagree with Indonesian claims that, unlike typical concessionary arrangements, oil exploitation in Indonesia is in the hands of Indonesians; yet, an examination of existing relationships between Pertamina and the oil companies reveals the disparity between the written word and reality. The legal differences between production-sharing and concession contracts are often devoid of operational significance.

'In particular,... contractors have retained effective management control in spite of the management clause. The functions which contractors have so far performed are virtually indistinguishable from those performed by concessionaires having formal equity interest and exclusive management prerogatives....'

5. Even in the traditional concession, a host country continued to possess ownership and to exercise *ultimate* control over the resources under concession.

6. This discussion is mainly based on Albert T. Chandler, 'Mineral Exploration and Development: Some Basic Considerations/Trends in Government Management of Mineral Exploration and Development', UN, ESCAP Document No. E/ESCAP/NR.3/6, 21 July 1976, pp. 13–14.

7. The term 'bonus' applies to a cash payment made at the time of acquiring a lease on a property (lease bonus) or upon granting or signing a work contract or concession (signature bonus). Payments may also be made at various stages of the exploration and production process, e.g., a discovery bonus or production bonus. See Roderick O'Brien, *South China Sea Oil: Two Problems of Ownership and Development*,

Occasional Paper No. 47 (Singapore: Institute of Southeast Asian Studies, 1977), p. 2, and Megill, *Exploration Economics*, p. 23.

8. The term 'rentals' refers to payments for surface area leased or for concessions granted (O'Brien, op. cit., p. 2).

9. The term 'royalty' normally refers to a compensation or portion of the proceeds paid to the owner of a right, as in oil or mineral rights. (Random House, *College Dictionary*.) In the petroleum industry, royalties may be payable either as a percentage of production or as a fixed sum per barrel output. The oil industry defines *royalty* as a 'share of the gross production of oil and gas on a property of the landowner without bearing any of the costs of producing the oil or gas'. See R. D. Langenkamp, *Handbook of Oil Industry Terms and Phrases* (Tulsa, Oklahoma: The Petroleum Publishing Company, 1974).

10. See Fabrikant, op. cit.

11. This paragraph is based on the article, 'From Concessions to Contracts-II: End of an Era', appearing in *PE*, January 1975, pp. 21–4.

12. See *PE*, January 1975, op. cit., for a well-written summary of the global developments.

13. The following summary is based largely on Chandler, op. cit., and O'Brien, op. cit.

14. See Chandler, op. cit., p. 16.

15 See, for example, O'Brien, op. cit., p. 63.

16. See ibid., p. 63.

17. Source for this section: State of Brunei, Enactment No. 3, 1963; Enactment No. 6, 1963; Enactment No. 5, 1969; Enactment No. 5, 1972.

18. *PE*, December 1978, pp. 505–6, and private consultant, Singapore.

19. UN, ESCAP, MRDS No. 10, 1959, and *PN*, January 1976.

20. At the time of writing, this figure which appears out of line (in view of the area limit cited for the mining lease) could not be verified.

21. *PN*, January 1975, p. 52.

22. *PN*, January 1976, and private source. See also *PE*, December 1978, pp. 505–6.

23 Three-quarters of each contract area must be relinquished upon commercial discovery. See *PE*, December 1978, p. 505.

24. 'Commercial discovery' was defined as 'an accumulation of petroleum which the state company and the contractor decide to develop'. (Information obtained from a private source.)

25. Private source.

26. In earlier arrangements, all sales proceeds from exports of production were

required to be remitted back to the Union Bank of Burma, and cost recovery and profit allowances were then to be transferred to the contractor's bank abroad. This gave the contract an element of 'profit-sharing' rather than 'production-sharing'. Negotiations covering this aspect, to reduce problems related to the rate of foreign exchange used in such transactions, resulted in stipulations that crude oil *production* rather than proceeds would be shared. (This information was obtained from a private source.)

27. Also earlier known as the Khmer Republic. This section is based on information from *PN*, January 1977, p. 11, and UN, ESCAP, MRDS No. 40, pp. 100–1, 128–9.

28. Source: UN, ESCAP, MRDS 40, pp. 94–6, UNDP/CCOP, *Report of the Sixth Session*, p. 136, and actual copies of the basic legislation. The information on legislation dates in parentheses were supplied by the Directorate General of Oil and Gas.

29. Source: Bartlett, op. cit., pp. 194–5. See also Fabrikant, op. cit., 1973(a). Additional information was received in conversations with Indonesian government officials, 29 May 1978.

30. This affected Stanvac's acreage only.

31. This only covered the acreage of Caltex.

32. Source: Pertamina officials. Remuneration to the contractor is computed at costs plus fee (fee is 20 U.S. cents).

33. Production and compensation bonuses are not included in computing operating costs under the production sharing contract. Although not explicitly stated in the contract, signature bonuses appear to fall under the same exclusion. This was confirmed by the Legal Division of MIGAS.

34. See Section 14.9 in the sample production-sharing contract in Bartlett, op. cit., p. 360, on this provision and on the method of calculating this portion. See also Appendix B, Section V(1.2(p)).

35. The information in the following was basically as explained in Sritua Arief, *Financial Analysis of the Indonesian Petroleum Industry* (Jakarta: Sritua Arief Associates, 1977), pp. 164–228.

Unclear aspects were discussed and rechecked with the Indonesian government in May 1978. See also the section on developments in Indonesia in Chapter III for the reasons for some of the changes in format and conditions.

36. This was, however, not final, according to GOI officials, but was still under negotiation in May 1978. Nothing official has appeared in the press by the time this was being written.

37. Private source and confirmed by Indonesian government. See also U.S., Internal Revenue Ruling 78–222.

38. See section in Chapter III on developments in Indonesia's production-sharing contracts for the definition of the classifications *Group I* and *Group II*.

39. As explained earlier, taxes are at the rate of 56 per cent on the company's share, broken down into the following: 45 per cent corporate tax and 11 per cent dividend tax. The computation of the 65.91/34.09 split is arrived at as follows (with the 15 per cent net company profit share already pre-determined):

$$Q_c (1.00-0.56) = .15 (Q_t) \qquad \text{where } Q_t = \text{total 'profit oil'}$$
$$Q_c = \text{company's share before taxes}$$

or $.44Q_c = .15Q_t$

40. This was confirmed in a conversation with an Indonesian government official. The current rate was 22.06 per cent off the company's 15 per cent 'net' profit oil.

41. The classification is arbitrary and follows nomenclature in the press. In essence, however, the arrangement is not considered any different from normal partnership arrangements other than that the partner is the state enterprise rather than another private oil company. See *AWSJ*, 6 July 1977, p. 3, and other press accounts during the period. These were verified in conversations with Indonesian government officials and oil company executives in May 1978 in Jakarta.

42. This section is basically from UN, ESCAP, MRDS No. 40, p. 129.

43. Information released by Petronas (see their Report), supplemented by information given in *PN*, January 1978, pp. 30–2, and *FEER*, 8 July 1977, pp. 36–7.

44. After the 15-year production period, any remaining reserves revert to the government. (Private source.)

45. 1 kilolitre = 6.2898 barrels.

46. From Petronas, July 1978.

47. See Act 95, Petroleum Mining Act, 1966, revised 1972, and details of concession terms summarized in *PN*, January 1974, p. 40.

48. One hectare = 0.01 square kilometre.

49. *PN*, January 1976, p. 54, supplemented by government and industry sources in 1978.

50. Government official.

51. Government and industry sources, 1978.

52. *PN*, January 1976, p. 54. Originally, P.D. 87 only allowed for 40 per cent maximum cost recovery. See Section 8(1) (2).

53. Government and industry sources.

54. P.D. 1206, Section 12(2). See also *PN*, January 1978, p. 40.

55. *PN*, January 1976, and government source.

56. *PN*, December 1978, Supplement.

57. References to these will henceforth be PA for the Petroleum Act, and PITA for the Petroleum Income Tax Act.

58. See 'Proclamation on Demarcation of the Continental Shelf of Thailand in the Gulf of Thailand'. Thailand ratified the Geneva Convention on the Continental Shelf on 2 July 1968.

59. UNDP/CCOP, *Technical Bulletin*, Vol. 11, p. 144.

60. See *Ministerial Regulations Issued under the Provisions of the Petroleum Act and the Petroleum Income Tax Act* (Bangkok: Department of Mineral Resources, July 1973).

61. See Thai Delegation, 'A General Account of Petroleum Legislation and Policies in Thailand', document I&NR/PL/C.R., Paper 17, 27 September 1971, in UN, ESCAP, MRDS No. 40, p. 118. This paper did not elaborate on what those 'special advantages' were.

62. Cited in *PN*, January 1978, p. 46.

63. This was reproduced in full in *AWSJ*, 7 June 1977, p. 3.

64. The *PN* report included Elf Aquitaine among those receiving awards from the government of Vietnam in 1977. A later article in *PE*, February 1979, noted that the 1977 protocol signed by Elf for two blocks had fallen through (see 'Problems loom for renewed search', p. 63).

65. See *PE*, February 1979, p. 63. The *PE* report on these provisions differs slightly from earlier reports. In *PN*, January 1978, p. 46, for example, the 'buy-up' privileges was indicated to be at 42 per cent of production, at 7–10 per cent below the prevailing international price. The *AWSJ* article, 19 April 1978, p. 3, reported such buy-up privileges to be at 'market prices'.

66. See *PE*, February 1979, p. 63.

67. See Tim Williams, 'Vietnam wants American oil firms', in *PN*, August 1978, pp. 8, 10.

68. *PE*, May 1978, p. 212, and Bow Valley press release, 6 September 1978 (Calgary, Canada).

69. *PN*, January 1974, p. 47.

70. The outline is based on information in UN, ESCAP, MRDS, No. 40, pp. 135–6, and *PN*, January 1974, pp. 42–5, and April 1973, pp. 28–9.

71. UN, ESCAP, MRDS, No. 40, p. 135.

72. *PN*, January 1974, pp. 42 and 44.

73. *PN*, January 1974, p. 43.

74. *PN*, January 1974, p. 45.

V

The Supply of Reserves: Implications of Institutional Arrangements

THE preceding chapters have explained the following: (1) the basic assumptions in this study on the strategies that might be expected of the agent supplying petroleum reserves, (2) the fundamental characteristics of the petroleum industry in general that distinguish it from others as well as the special features of the market in South-East Asia, and (3) the overall property rights or contractual environment within which the petroleum exploration firm operates in South-East Asia.

In this chapter we will look more closely at the overall contractual framework and attempt to understand the implications of specific aspects of this framework for the supply of petroleum reserves, in the light of our assumptions on firm behaviour and of our understanding of the industry's structure and technical characteristics. Because the Indonesian industry has a longer and more intensive history, discussion will tend to refer to the Indonesian case more frequently than to other countries in South-East Asia.

Fiscal Framework and Exploration Incentives

For purposes of illustration, Table 5.1 summarizes the basic terms of the production-sharing contracts which the Indonesian government has concluded since it introduced the system in 1965 up to the end of 1977.

There were several reasons for putting the tedious details of this table together. The primary reason, however, was that it appeared

108

TABLE 5.1

PRODUCTION-SHARING CONTRACTS IN OPERATION BETWEEN 1975 AND 1977 AND TERMS OF ORIGINAL CONTRACT

Signing of Contract Year/Mo.	Contractors*	% Share in Oper. as of 1/1/78	Original Contract Area (km²) Onshore	Offshore	Signature & Compensation Bonuses Total $10⁶	$/km²	Minimum Expenditure $10⁶	$/km²	Company's Original Production Share** Basic %	% @ XMBD	Production Bonus $10⁶ @ 10³ b/d Level 1	Level 2 and Others	Revised P/S 1977
(1)	(2)	(3)	(4)	(5)	(6)	(7)	(8)	(9)	(10)	(11)	(12)	(13)	(14)
1961/Sept. 1	Asamera	60.0	3 390	—	—	—	7.50	2 212	40	—	—	—	85–15
1964/Mar. 10	Indonesian Gulf (ex. Refican)	50.0	23 000	3 499 (1965)	—	—	10.00	377	35	—	—	—	—
1966/Aug. 18	IIAPCO (ARCO)	46.0	—	54 778	—	—	7.50	137	35	—	—	—	85–15
1966/Oct. 6	JAPEX	?	—	32 490	—	—	7.50	231	35	—	—	—	85–15
1966/Nov. 22	Kyushu Oil (T)	15.66	—	49 857	—	—	25.75	516	35	—	5 @ 50	10 @ 100	—
1966/Nov. 26	Asamera (?)* (or Mobil? ex Asamera)	— (100)	3 745	—	1.0	267.0	7.50	2,002	35	—	—	—	—
1967/Apr. 1	Cities	33.3	—	154 160	—	—	7.50	49	35	—	1 @ 40	4 @ 50	85–15
1967/May 12	Continental Oil	12.5	16 580	—	1.0	60	12.00	724	35	32.5 @ 75	3 @ 75	—	—
1968/Jan. 26	Union Oil	100.0	135 223	—	0.65	5	3.20	24	35	32.5 @ 50	—	—	85–15
1968/Apr. 8	International Oil (T)	100.0	25 885	(Partial)	—	—	2.45	95	35	32.5 @ 75	1 @ 50	—	—
1968/May 28	Phillips/Superior	Sup (20) Phillips (5)	311 505	(Partial)	0.5	2	17.00	55	35	32.5 @ 75	2 @ 100	—	—

Signing of Contract Year/Mo.	Contractors*	% Share in Oper as of 1/1/78	Original Contract Area (km²)		Signature & Compensation Bonuses		Minimum Expenditure		Company's Original Production Share**		Production Bonus $10⁶ @ 10³ b/d		Revised P/S 1977
			Onshore	Offshore	Total $10⁶	$/km²	$10⁶	$/km²	Basic %	% @ XMBD	Level 1	Level 2 and Others	
(1)	(2)	(3)	(4)	(5)	(6)	(7)	(8)	(9)	(10)	(11)	(12)	(13)	(14)
1968/July 6	Total (T)	50.0	21 483	—	1.0	47	10.5	489	35	32.5 @ 75	1 @ 100	2 @ 200	—
1968/Aug. 8	Huffington	10.0	11 920 + 5 045	—	1.0	59	11.0	648	35	32.5 @ 100	1 @ 100	—	85–15
1968/Sept. 6	IIAPCO	52.68	—	123 995	1.25	10	22.5	182	35	32.5 @ 75	3.5 @ 75	—	—
1968/Oct. 10	AGIP/SPA	50.0	100 535	—	1.5	15	16.0	159	35	32.5 @ 75	5 @ 75	1.5 @ 100 2 @ 200	—
1968/Oct. 15	REDCO (per Migas list)	100.0	582 + 285 + 11 256 = 12 123	—	—	—	5.5	454	35	—	—	—	85–15
1968/Nov. 6	Total (ex Java Sea)	50.0	—	9 494	—	—	0.625	66	35	—	—	—	—
1968/Oct. 16	Continental Oil	40	—	108 705	7.0	68	14.0	135	35	32.5 @ 75	3 @ 50	3 @ 100	85–15
1968/Oct. 16	Mobil	100.0	—	31 995	1.25	39	5.5	172	35	—	5 @ 50	5 @ 100 5 @ 200 5 @ 300	—

TABLE 5.1 (continued)

Signing of Contract Year/Mo.	Contractors*	% Share in Oper. as of 1/1/78	Original Contract Area (km²)		Signature & Compensation Bonuses		Minimum Expenditure		Company's Original Production Share**		Production Bonus $10⁶ @ 10³ b/d		Revised P/S 1977
			Onshore	Offshore	Total $10⁶	$/km²	$10⁶	$/km²	Basic %	% @ XMBD	Level 1	Level 2 and Others	
(1)	(2)	(3)	(4)	(5)	(6)	(7)	(8)	(9)	(10)	(11)	(12)	(13)	(14)
1968/Oct. 25	Union Oil	100.0	—	24 485	0.425	17	4.0	164	35	32.5 @ 75	1.5 @ 75	1.5 @ 175	85–15
1968/Dec. 19	AGIP/SPA	33.33	—	104 620 (Partial)	1.5	14	21.0	201	35	32.5 @ 75	0.5 @ 50	1 @ 75, 2 @ 100, 2 @ 200	85–15
1969/Aug. 9	White Shield (ex Asia Oil(?))	95.5	72 500	Partial	2.75	38	17.5	241	35	32.5 @ 75	1 @ 50	5 @ 200	85–15
1969/Nov. 1	Associated Aust. Resource (ex Gulf/Western)	100.0	77 700	Partial	3.0	39	14.5	187	35	32.5 @ 55	1.5 @ 25	3 @ 50, 5 @ 100	85–15
1969/Dec. 6	Kaltim Shell N.V	50.0	32 060	—	5.0	156	21.28	664	35	32.5 @ 75	1 @ 50	1 @ 100	85–15
1969/Dec. 9	Total (T)	50.0	14 000	—	2.0	143	10.0	714	35	32.5 @ 75	1 @ 50	1 @ 100	—
1970/Feb.4	Tesoro (ex W. Phillips) (T)	50.0	31 990	Partial	0.5	16	17.5	547	35	32.5 @ 75	1 @ 50	1 @ 75, 1 @ 100	—
1970/Feb. 9	Calasiatic/Topco (Amoseas) (T)	100.0	—	56 665	2.0	35	9.8	173	35	32.5 @ 65	1.5 @ 50	2 @ 100	—
1970/Mar. 2	BP Petroleum Dev.	25.0	27 920	Partial	0.75	27	9.15	328	35	32.5 @ 75	1 @ 50	2 @ 75	85–15

Signing of Contract Year/Mo.	Contractors*	% Share in Oper. as of 1/1/78	Original Contract Area (km²) Onshore	Offshore	Signature & Compensation Bonuses Total $10⁶	$/ km²	Minimum Expenditure $10⁶	$/km²	Company's Original Production Share*** Basic %	% @ XMBD	Production Bonus $10⁶ @ 10³b/d Level 1	Level 2 and Others	Revised P/S 1977
(1)	(2)	(3)	(4)	(5)	(6)	(7)	(8)	(9)	(10)	(11)	(12)	(13)	(14)
1970/Aug. 5	Pan Ocean (ex Kondur)	35.0	—	39 550	4.0	101	11.0	278	35	32.5 @ 75	1 @ 50	2 @ 100	—
1970/Oct. 15	Petromer-Trend	27.0	5 000	—	0.15	30	5.35	1,070	35	32.5 @ 75	1 @ 50	2 @ 100	85–15
1970/Oct. 24	Indonesian Gulf	NA	10 000 ±	Partial	1.0	100	8.5	85C	35	32.5 @ 75	1 @ 50	1 @ 100	—
1970/Oct. 24	Louisiana Land Expl. (ex Whitestone)	15.0	14 997	—	3.0	200	10.15	677	35	32.5 @ 75	0.5 @ 50	1 @ 75, 1 @ 100, 2 @ 200	85–15
1970/Nov. 28	Asia Oil (ex Pexa Oil) (T)	7.5	1 489	—	0.25	168	7.25	4,869	26.25	24.375 @ 75	0.75 on continuous production for 5 years	—	—
1970/Jan. 15	Djawa Shell (T)	75.0	—	9 000	4.0	444	18.0	2,000	35	32.5 @ 75, 30 @ 200	1 @ 50	1 @ 75	—
1971/Aug. 9	ARCO	25.0	16 460	—	5.0	304	19.75	1,200	35	32.5 @ 50, 30 @ 75	2 @ 50	4 @ 100	—
1971/Aug. 9 (eff. Nov. 28/83)	Caltex	100.0	16 505	—	8.0 / 8.0	485 / 320	36.0 / 15.0	2,181 / 601	30	—	—	—	—
1971/Aug. 9	Calasiatic/ Topco (T)	100.0	24 975	—					35	32.5 @ 60, 30 @ 100	2 @ 60	3 @ 100	85–15
1971/Oct. 28	Continental Oil	40.0	52 747	—	2.0	38	17.5	332	35	32.5 @ 60, 30 @ 100	1 @ 50	2 @ 75, 4 @ 200	—

TABLE 5.1 (continued)

Signing of Contract Year/Mo. (1)	Contractors* (2)	% Share in Oper. as of 1/1/78 (3)	Original Contract Area (km²)		Signature & Compensation Bonuses		Minimum Expenditure		Company's Original Production Share***		Production Bonus $10⁶ @ 10³ b/d		Revised P/S 1977
			Onshore (4)	Offshore (5)	Total $10⁶ (6)	$/ km² (7)	$10⁶ (8)	$/km² (9)	Basic % (10)	% @ XMBD (11)	Level 1 (12)	Level 2 and Others (13)	(14)
1972/Mar. 3	Indonesia Offshore (Champlin) (T)	53.8	—	74 074	0.75	10	13.25	179	35	32.5 @ 60 30 @ 75	1 @ 50	1 @ 75	—
1972/July 27	Total (T)	50.0	3 150	—	1.0	318	3.0	952	35	30 @ 50	1.5 @ 50	—	—
1973/Mar. 14	Mobil	100.0	—	20 000	1.375	69	9.0	450	35	32.5 @ 100 30 @ 150	2 @ 100	3 @ 150	—
1973/Oct. 6	Sun Oil	100.0	17 760	—	2.5	125	11.35	568	35	32.5 @ 40 30 @ 75	0.5 @ start of commercial production	1 @ 50 1 @ 100 1 @ 200	—
1974/Jan. 14	Katy Industries and Union Oil	20.0 80.0	23 670	—	10.0	423	17.5	739	30	25 @ 50 20 @ 300	1 @ 50	1 @ 100 1 @ 150	—
1975/Jan. 16	Continental Oil	50.0	3 865	3 865	2.0	518	17.1	4,424	30	20	2 @ 50	5 @ 100 7 @ 250 10 @ 500	85–15
1975/Jan. 20	Calasiatic/Topco	100.0	6 865	—	8.0	1,165	10.0	1,457	30	25 @ 50 20 @ 100	1 @ 50	1 @ 100	—
1975/Mar. 20	Moncrief Pexpac	42.0	—	29 390	0.75	26	11.85	403	30	25 @ 50 20 @ >125	1.5 @ commercial production	—	—

Signing of Contract Year/Mo.	Contractors*	% Share in Oper. as of 1/1/78	Original Contract Area (km²)		Signature & Compensation Bonuses		Minimum Expenditure		Company's Original Production Share**		Production Bonus $10⁶ @ 10³ b/d		Revised P/S 1977
			Onshore	Offshore	Total $10⁶	$/km²	$10⁶	$/km²	Basic %	% @ XMBD	Level 1	Level 2 and Others	
(1)	(2)	(3)	(4)	(5)	(6)	(7)	(8)	(9)	(10)	(11)	(12)	(13)	(14)
1975/Mar. 20	INCA Ltd.	100.0	—	20 030	1.25	62	10.2	509	27.5	22.5 @ 50 20 @ >150 15 @ >300 10 @ >500	1.5 @ commercial discovery	1 @ 50 1 @ 75 2 @ 100	—
1975/Mar. 22	Phillips Tenneco	33.4 33.4		5 165	6.0	1,162	20.5	3,969	27.5ᵇ	22.5 @ >50 20 @ 150	2 @ 50	2.5 @ 100 3 @ 200	—
1975/July 15	INCA Ltd.	100.0		5 955	7.5	1,260	10.5		27.5	22.5 @ 50 20 @ 150 15 @ 300 10 @ 500	1 @ 50	1.5 @ 75 2 @ 100	—
1977/Oct. 22	Continental Oil	50.0	9 200	—	3.15	342	40.6 (drilling of 1 well in year 1)	441	15ᵃ	—	1 @ 50	1 @ 150	—
1977/Oct. 22	Total	50.0	8 890?	—	6.0		25 (1 well in year 1)		15ᵃ	—	2 @ 100	2 @ 200	—

ᵃ34.09 per cent gross but equivalent to 15 per cent net, in comparable terms with earlier contracts.

ᵇCosts recovery allowance @ 35 per cent.

T = Terminated at time of writing.

Sources: Indonesia, Directorate General of Oil and Gas; Petroleum News, January 1976, 1977, 1978 issues; Situa Arief, The Indonesian Petroleum Industry, 1976.

*As listed in MIGAS list.

**Contracts were renegotiated in 1976.

desirable to see if there was any meaningful pattern in the contract terms, but especially in the 'front-end' investment requirements (the signature bonuses and minimum expenditure requirements), that might be related to the exploration investment behaviour of the firms.

THE EXPENDITURE PROVISIONS

Columns 7 and 9 show the cost in U.S. dollars per square kilometre of contract area of initial (signature and compensation)[1] bonuses and of minimum expenditure requirements over specific years during the exploration period. It is clear from a visual inspection that, although there are some patterns in absolute amounts that may be related to the type of expenditure and the period during which the contract was made (e.g., minimum expenditure requirements in 1966 were the same), expenditures required were in no way related to contract area size.

It is equally difficult to attempt to draw any conclusions on the pattern followed by production bonuses. Columns 12 and 13 tend to support the statements made by government officials that these terms result from negotiations. An Indonesian official indicated that beginning in 1967 contracts were awarded on the basis of tenders, and that the government awarded a contract to a company offering the terms most favourable to the Indonesian government.[2] The levels and sizes of the bonuses may therefore indicate a company's perceptions of an area's petroleum prospects.

The production shares tend to present a more consistent picture over the period 1967 to 1975. With only a few exceptions, the company's share of 'profit oil' was reduced to 32.5 per cent at production levels averaging 75,000 barrels per day, beginning with the contract of Continental Oil of 12 May 1967. A third level was introduced with the contract of Djawa Shell of 15 January 1971, where the company's share was further reduced to 30 per cent at production levels averaging 200,000 barrels per day. Subsequent contracts beyond this date, however, ceased to follow a pattern based on the output level.

Several statistical attempts were made in this study to determine some explanation for the size and/or frequencies of the two types of bonuses. The explanatory variables used in several combinations were: (1) the payment of bonuses at any stage of the total development cycle (signature, or production bonus); (2) the location of the contract area (offshore and onshore); (3) the size of the contracting company (whether major or not major)[3]; and (4) the average field size of discoveries.

The results of these tests were extremely poor. The low proportions[4] of the variations in the values of the dependent variable that were explained by the independent variables listed above tended to support the initial impression that, most likely, the contract terms were essentially the result of bargaining and were, therefore, indeterminate on an *a priori* basis.

COMPANY 'TAKE' AND EXPLORATION

Hefty increases in petroleum prices have provided the required incentive for exploring and developing South-East Asia's relatively more costly areas. Minas crude (more frequently referred to as 'Sumatran light') sold for only US$1.60–1.67 in early 1971, was selling at US$10.80 in early 1974, and began selling at US$13.90 on 1 January 1979 (see Table 5.2). In 1978 company profit margins per barrel were reportedly around US$1.00 per barrel. Although profit margins of the companies did not rise in line with price increases, these margins—which do not include company headquarters overhead expenses—appear to be substantial compared to the 25-cent margins received in the Middle East in 1975.[5] The Indonesian fields are much smaller than those in Saudi Arabia, however, and some of these Indonesian profits would be expected to find their way back into exploration investments.[6] The size of a contractor's production also makes a difference; e.g., Caltex's output versus the output of Cities Service.[7]

Industry analyses usually point to the number of new contracts as an indicator of oil company response to 'pre-acquisition' economic conditions, or the 'contractual terms'. Table 5.3 summarizes some of the information in Table 5.1. It will be noted that Indonesia's best years were 1968 to 1970. During these three years, Indonesia signed 32 PSCs, with half of them concluded in 1968. Over two-thirds of the contracts covered offshore areas, in whole or

TABLE 5.2

INDONESIAN CRUDE: F.O.B. PRICE PER BARREL AND REPORTED COMPANY COSTS AND MARGINS

As of	Sumatran Light* f.o.b.	Company Costs	Profit Margin
1st Quarter 1971	US$1.60–1.67[a]	—	—
10 Jan. 1975	US$12.80[a]	US$10.65[b]	US$2.15[b]
1 Jan. 1976	US$12.80[a]	US$11.65[b]	US$1.15[b]
1 Jan. 1978	US$13.55	US$12.50[b]	US$1.05[c]
1 Jan. 1979	US$13.90[b]	—	—

[a] Price data = Indonesia, Department of Mines.

[b] *Petroleum Economist*, June 1976, December 1976, February 1979.

[c] Based on profit margins estimated by industry. See *Asian Wall Street Journal*, 7 February 1977, 'Indonesia Plans New Incentives for Oil Firms', and hypothetical estimates appearing in U.S. Embassy, *Indonesia's Petroleum Sector*, July 1977 and July 1978.

Note: Minas crude is the representative crude, or the price-reference ('marker') crude in Indonesia. It has an API gravity of 35°. (See Appendix A.)

in part, with the total amount of offshore area covered by contracts reaching a peak in 1971. It might be interesting to note at this point that most of the PSCs are held by relatively small companies[8] or independent contractors, and their willingness to enter into contracts in the offshore areas—where exploration is more costly than on land—may indicate that the contract terms were reasonable.[9] Despite increasing oil prices in the 1970s, including the oil embargo of 1973, however, the number of new contracts from 1972 to 1974 were at close to their lowest level of zero in 1976. It might be worth noting that the Indonesian government introduced its third level of government 'take' of 70 per cent of 'profit oil' with Shell's contract of 1971 (see Table 5.1). The only visible reaction of the companies at this point to the fourfold price increase of 1973–4 appeared to be in the increased drilling activity on existing contract areas (see Figure 6.2). The apparent lack of interest in 1972–4 was followed by a slight upsurge of seven new contracts in 1975 (six of which were offshore), despite the more stringent pro-

TABLE 5.3

SUMMARY OF NEW INDONESIAN PRODUCTION-SHARING
CONTRACTS TO 1977

Year	New Contracts	Offshore	
		New Contracts (All or Part)	Cumulative Area (10^6 km^2)
	(1)	(2)	(3)
1961	1	0	0
1964	1	1	NA
1966	4	4	NA
1967	2	1	0.39
1968	16	14	1.68
1969	8	6	2.04
1970	9	6	2.86
1971	4	0	2.86
1972	2	1	2.23*
1973	3	1	1.74
1974	1	0	1.23
1975	7	6	1.22
1976	0	—	NA
1977	2	0	NA

Sources: Columns 1 and 2, from Table 5.1.
  Column 3 from UNDP/CCOP, *Technical Bulletin*, Vol. 11, Table 2A.

  *Mandatory relinquishment started.
  NA = Not available.

duction-sharing conditions. (The Indonesian government had introduced the 30–25–20 company-take provision in 1974 and continued this in 1975, along with contracts containing company-'take' provisions on a declining basis of 27.5–22.5–20–15–10.) Even with this upsurge, however, total offshore contract areas had, after mandatory partial relinquishments and voluntary total relinquishments, been reduced to a level lower than that of 1968 (see Table 5.3 again).

The complete standstill in 1976 in new offers from companies plus the drastic downturn in exploratory drilling (see Figure 6.2) may be attributed generally to the Pertamina financial crisis and to subsequent contract revisions to provide the host government with additional and immediate cash flow.[10] Companies in a host country do not, after all, operate in a vacuum, but as multinationals are able to rank their prospects over several geographical regions. A review of the record in the Philippines during the period 1972 (when service contracts were first offered) to 1977 may provide some insights (see Table 5.4). The number of new contracts in the Philippines during the period, viewed both in the light of the absence of any significant discovery until mid-1977 and the downturn in Indonesia, may be an indication of the investors' perception both of the geological risk as well as of prospective returns.

Relating this to the theoretical discussion in Chapter III, the effect of increased government 'take' may be viewed in terms of its impacts on the economic limit of a petroleum reservoir. In the earlier discussion, it was pointed out that not all of the petroleum in a reservoir would be recovered because, given the geologic conditions and technical aspects of producing oil, at some point the marginal cost of producing a barrel would exceed the marginal revenue that would be received from selling that barrel of oil. That is:

Economic limit = 0 = Marginal Revenue – Marginal Cost

TABLE 5.4
PHILIPPINES: NEW SERVICE CONTRACTS, 1972–1977

Year	New Contracts	No. Foreign[b]
1972	1	1
1973	8	8
1974	3	3
1975	2	2
1976	10[a]	1
1977	5[a]	4

Source: Philippine Government, Bureau of Energy Development.

[a] Includes Geophysical Survey Contracts in 1976 and 1977 which are convertible to regular service contracts.

[b] Refers only to principal contractor. Contractors usually enter into partnerships; local companies, in particular, normally have foreign partners.

If the technical costs (that is, costs excluding government 'take') are unchanged and assuming the risk discount factor remains unchanged, the economic limit of the reservoir will be reached much earlier if government 'take' increases marginal cost.[11] This limit may indicate to a company calculating its pre-acquisition costs that certain prospects considered commercially viable under one set of conditions may not be viable under a different set of conditions.[12]

In this connection, it might be useful to note the implications of the variable profit-split provisions and production bonuses that have become standard in South-East Asian production-sharing contracts in the 1970s. (See Chapter IV again.) On one hand, forms of contingent payments[13] incorporated in a production-sharing contract during negotiations may be viewed as transferring part of the risks of exploration and production from the private firm to the government. On the other hand, this may be viewed as decreasing the company's share of output progressively with increasing production levels; thus, the contract in effect imposes a penalty for increasing production, or, seen another way, for discovering fields of increasingly larger sizes. Such penalty-reward arrangements would appear to have an influence in field-size estimation. In the same way, by penalizing a company for higher production levels, the production bonus reduces the estimated economic limit of the reservoir.

Comparative analysis of company 'take'. Table 5.5 shows a comparative calculation of the effective share of companies per 100 barrels of output in Indonesia, Malaysia, and the Philippines under past and existing contract conditions at the time of writing. The 'effective' company share includes cost recovery oil but excludes taxes; it also does not include any production bonuses the company may pay beyond a certain production level. This means that a company's net share of all output is actually lower than as shown at output levels of 50,000 or 75,000 barrels per day, and even smaller at higher levels for the Indonesian cases.

The table shows that the Philippine contracts provide the largest effective company 'take' per 100 barrels of the three countries compared. The exception is the short-term situation for a small field in

TABLE 5.5
COMPARISON OF COMPANY 'TAKE' PER 100 BARRELS OF OIL UNDER DIFFERING CONTRACTUAL ARRANGEMENTS

| | Indonesia | | | Malaysia | Philippines | |
	Pre-1976	Post-1976(I)[a]	Post-1976(II)[b]	1976	1972 Contracts	1975 Contracts
1. Base of 100 barrels	100	100	100	100	100	100
2. Cost recovery allowance	40	40	120	20	70	55
3. Royalties	—	—	—	10	—	—
4. Balance—'profit oil'	60	60	−20 (deferred)	70	30	45
5. Effective company share of 'profit oil'[c]						
@ 30.0 per cent				21		
@ 34.09 per cent		20.45				
@ 35.0 per cent	21					15.75
@ 40.0 per cent					12	
6. Add: Tax allowance	27	—	—	—	—	—
7. Add: Cost recovery allowance	40	40	100	20	70	55
8. Pre-tax company share	88	60.45	100	41	82	70.75

	Indonesia			Malaysia	Philippines	
	Pre-1976	*Post-1976(I)*[a]	*Post-1976(II)*[b]	*1976*	*1972 Contracts*	*1975 Contracts*
9. Less: Tax oil on (5)						
@ 35 per cent	—	—	—	—	4.2	5.5
@ 45 per cent	—	—	—	9.45	—	—
@ 56 per cent	27	11.45	—	—	—	—
10. Post-tax company share	61	49	100	31.55	77.8	65.25
11. Less: domestic pro rata @ remuneration of 20 ¢/bbl.	4.6[d]	2	—	—	—[e]	—[e]
12. Effective net company 'take' (before production bonus)	56.4	47	100	31.55	77.8	65.25
13. Government's share (pre-production bonus)	43.6	53	0	68.45	22.2	34.75

[a] Assumes the double declining method results in recovery allowance of 40 per cent in the area.
[b] Assumes cost recovery exceeds output in area.
[c] 'Effective' share refers to actual share relative to total output.
[d] Computed at 22.06 per cent of company's share of 'profit oil'.
[e] Domestic allocation is sold to the government at world market prices.

Indonesia, where cost recovery might exceed output—what we have called 'post-1976 (II)'. Malaysia's contract terms are the most stringent of the three.

'Risk' Trade-offs

Companies will, of course, weigh the geologic opportunities as well as the contractual factors. Companies may opt for the more attractive geologic prospect which it knows enough about and where it perceives its risks to be smaller, rather than choose another area where the contractual conditions appear to be better but where the risks are much higher. On the other hand, depending on its trade-offs, it may do the opposite.[14] The alternatives may be intra-regional or inter-regional. For example, the company's trade-offs may be between the Philippines versus Malaysia, or Indonesia versus the United States' Alaska or Atlantic offshore areas.

COMPANY SIZE AND RISK

Success in exploration, by definition, implies huge outlays of capital funds for the development of producing capacity. (See Chapter III again.) Even before arriving at the development stage, however, a substantial amount of money is required. Table 5.3 shows that even for Indonesia most new exploration areas were located offshore. The first commercial fields discovered in the Philippines were offshore, as are the major discoveries in Malaysia. Offshore exploration, where South-East Asia's best potentials lie, requires larger outlays than onshore search. As a general rule, because of the higher risks involved at this stage, exploration by private firms is financed by equity capital or internally generated funds. The cost recovery provisions of production-sharing contracts, which require companies to bear all the risks of exploration, further suggest the need to rely on internally generated funds. Table 5.6 presents the sources of funds for major oil companies active in exploration in Asia during the years 1972 to 1976. It shows that over two-thirds of the funds of the major oil companies covered by Copp's study were internally generated.

It would thus appear at first glance that only truly large com-

TABLE 5.6
CONSOLIDATED SOURCES AND USES OF FUNDS FOR MAJOR OIL COMPANIES ACTIVE IN THE ASIAN AREA
(millions of dollars)

	1972	1973	1974	1975	1976
Sources of Funds					
Total funds from operations	12,922	17,956	24,480	18,243	22,522
Total external sources	2,773	2,571	4,243	6,354	5,951
Other sources of funds	2,076	4,224	7,752	3,813	3,618
Total sources of funds	17,771[a]	24,750[a]	36,474[a]	28,410[a]	32,091[a]
% from operations	72	73	67	64	70

Companies in consolidation:

British Petroleum Co. Ltd.	Texaco Inc.
Union Oil Co. of California	Atlantic Richfield Co.
Royal Dutch Petroleum Co.	Mobil Corp.
Standard Oil Co. (Indiana)	Continental Oil Co.
Tenneco Inc.	Getty Oil Co.
Standard Oil Co. (Calif.)	Natomas Co.
Cities Service Co.	Exxon Corp.
Asamera Oil Corp.	Gulf Oil Corp.
Sun Co.	Phillips Petroleum Co.
Tesoro Petroleum Corp.	

Source: Salomon Brothers, as presented in E. Anthony Copp, 'Capital Sources versus Capital Demands in Asian Petroleum Markets', paper presented at the 1978 Offshore South-East Asia Conference, Singapore, 21–24 February 1978.

[a] Details may not add up because of rounding.

panies should attempt to engage in exploration in the region. But actual observation indicates the presence of many relatively small companies and independent operators. These small companies have been able to reduce their total risks by spreading their investments in small shares in several contract areas; thus the loss on one unsuccessful area may be borne by three, four, or five contractors.[15] This has been the pattern for most countries in Asia, but is more pronounced in Indonesia and the Philippines.

POLITICAL RISKS AND THE DISCOUNT FACTOR

Adelman[16] estimated that exploration and development combined would require a return of about five times the basic interest rate for a small company unable to spread its risks, and probably half that, or three times the basic rate for a larger company. Actual returns on a 5,000-acre tract in the Gulf of Mexico were estimated by a U.S. Bureau of Mines group to range from 14 to 17 per cent, depending on the rate of extraction of oil and natural gas. The returns, however, were computed only on the successful sites; Logue suggests that when the unsuccessful tracts are included in the estimation, a downward adjustment might result in a rate of return closer to 8 per cent per annum as estimated by Bybee.[17]

'Political risks' or uncertainty can, however, alter the company's trade-off factors. Uncertainty can be increased by unilateral changes in government policies and legislation on very short notice, when such shifts alter the property rights arrangements. Whether favourable or unfavourable to the firm, such a change triggers a reaction from the party affected. If such change results in new restrictions or reductions in contractual claims to the share of returns from a venture, such attenuation of property rights affect the operator's expectations about his terms of trade. Short-term shifts thus create an additional uncertainty factor in the firm's estimation of the future value of its returns. Since the uncertainty factor is lower the shorter the time period involved, or the closer to the present the terminal date of the venture is, one choice the firm may make is to shorten the period within which to recover its investments and to increase its annual rate of return. The risk element in its discounting rate is thus increased, to accommodate this additional uncertainty (see earlier discussion on risks in Chapter III). Another choice may be to select an alternative site where the political risks are lower. A firm faced with worse problems of the same nature elsewhere may, on the other hand, rate South-East Asia higher along these lines.

An attempt to analyse how companies may adjust their rate of return under contractual arrangements to incorporate their perceived risks on investments in specific areas was presented at the 1978 Offshore Southeast Asia Conference.[18] (See Table 5.7.) Under certain hypothetical conditions, a given reservoir size discovery could generate risk-assessed values highly different from the no-risk rate of return guaranteed under contractual conditions. Thus, for example, although the Philippines may have a 'no-risk' per cent rate of return of 25 compared to 17 in the U.S. offshore, a company may prefer to invest its funds in the latter, given the relative risk-assessed values for the two prospects.

These hypothetical rates do not appear to be out of line with actual rates. The rate of return on investments negotiated by Union Oil Co. of Thailand on its gas discoveries on the Gulf was reported to be in the range of 18 to 20 per cent.[19]

COST RECOVERY AND THE RISK TRADE-OFF

A related question might be asked at this point. It will be recalled that the production-sharing contract basically allows for recov-

TABLE 5.7
ADJUSTING THE RATE OF RETURN FOR RISK ON A GIVEN SIZE DISCOVERY
(Hypothetical Cases)

Area	No Risk % ROR	Risk Assessed Relative Value*
Australia	28	21
Indonesia	15	22.5
Malaysia	4	12
Philippines	25	19
U.K. North Sea	20	30
U.S.A. Offshore	17	25.5

Source: Peter D. Gaffney and C. P. Moyes, Offshore Southeast Asia Conference, 1978, Figure 7.

ROR = Rate of return.

*Relative rate of return number after putting all countries on same technical and legislative base.

ery of costs associated only with the contract area where commercial production takes place, thus restricting the risk of dry holes to the contractor. As a result of the IRS ruling on U.S. companies, companies in Indonesia will, in certain instances, be allowed to recover costs incurred in unsuccessful blocks from production in successful blocks. This method will apply only to old contracts; in keeping with the Indonesian government's policy of restricting risk-bearing to the contractor, companies would be required to form new legal entities for each new contract area. (See Chapter III.) In essence, therefore, the choice variables facing a firm basically remain unchanged—that is, it knows that it can recover pre-production costs only if successful.

Since in general an entrepreneur may be considered to be risk-averse, the current situation would appear to motivate a firm to try harder than it would, under different contract conditions, to enter into contracts on areas where the probability of success is relatively high. This may account for the rather high success ratio for wildcats in Indonesia (see Table 6.1), even using the more conservative AAPG classifications for exploratory wells. The success ratio of cumulative drilling between 1970 and 1976 is close to one of four (24 per cent), with the ratio further rising between 1974 and 1976 to much greater than 1 : 4 (28 per cent). This is undoubtedly high by industry standards. The U.S. success rate of 16.88 per cent in 1976 was considered a 'high'; the success rate in 1975 was 14.35 per cent but up to 1974 it was only around 10 per cent.[20] The rising success rate in the U.S. is attributed to higher prices for petroleum that make even smaller fields profitable and worth completing. This could also be true for the small Indonesian fields, but the tendency of a firm to go mainly for the 'sure thing' may not be discounted. That is, the contract terms create a bias in favour of areas with high success probabilities.

An improvement in the quality of information possessed by a firm concerning a region reduces its perceived geological and technical risks. In very large fields such as the Middle East, knowledge accumulated from additional drilling may result in a downward revision of the geological and technical risk factor. This may not necessarily follow for areas with relatively smaller fields, where, as

Adelman puts it, the operator 'learns more and more about less and less'.[21]

External Factors

Factors external to the region or to host country conditions may, of course, influence the supply function for petroleum reserves. One of these was suggested earlier; it arises mainly from the international character of the companies involved, and relates to risk-spreading by these companies. Taxation policies in the home countries of the contractors also affect overseas investment activity. Internal cash-flow problems are a possible third factor.

It was suggested in Chapter III that companies operating in different nations may be doing so in order to reduce their risks of loss. Petroleum exploration companies, in particular, will be inclined to do this. Post-World War II developments in host country policies in the traditional producing countries have, no doubt, provided incentives to the major international oil companies to diversify their sources of crude petroleum and thus to explore for new fields in Asia and elsewhere. Other factors such as advances in technology and price increases have helped them follow through their initial efforts, but the need to diversify raw material sources is, nevertheless, an important investment decision element.

Tax policies in the mother countries are also important incentives for investing overseas. It has been suggested that current U.S. policy, for example, offers an implicit tax incentive to do so.[22] Horst notes that the incentive increases as the foreign income tax decreases, as less debt and more equity is used in financing foreign investment, and the lower the interest it pays on intra-firm debt. A concrete example of the role that U.S. taxation might have in influencing overseas petroleum exploration was the effect of the decision of the U.S. Internal Revenue Service in 1976 on Indonesia. This ruling disallowed a foreign tax credit to a U.S. oil company for production retained by the Indonesian government under the production-sharing contract.[23] Although one MNC oil company executive averred that this issue had not in any way influenced its acquisition decisions in two Asian countries,[24] it would be reasonable to expect that, where this U.S. tax ruling rendered a prospect commercially non-competitive, this ruling could be a marginal decision factor.

Internal cash-flow problems may, of course, affect individual company decisions, although the overall industry entry-and-exit total for any one year may balance out. One particular example was the decision of Atlantic Richfield in 1976 to retrench in Indonesia, because it needed its available funds. to develop its Alaskan finds upon resolution of the environmental issues that had held up such development. Given the more stable environment surrounding the Alaskan development project, both geologically and investment-wise, the ARCO decision was not at all unexpected. Similar cash-flow problems were given as a reason by a Petronas official for the loss of momentum in the negotiations with Continental Oil in Malaysia.[25]

1. The term 'compensation bonus' has been used to refer to payments made for data on the contract area provided by the host government.

2. Conversation in May 1978.

3. The definition of whether a contractor was a major oil company or not did not follow the conventional practice. For lack of convenient sources as a basis of classification, the companies covered in Table 5.5 and by the Chase Manhattan Bank in their financial analysis studies (often referred to as the Chase Manhattan 'Group of Petroleum Companies') or their subsidiaries were taken to be 'majors' and all others as not. The Chase companies are: Amerada Hess Corporation, Apco Oil Corporation, Ashland Oil, Inc., Atlantic Richfield Company, The British Petroleum Company Limited, Champlin Petroleum Company, Cities Service Company, Clark Oil & Refining Corporation, Compagnie Française des Petroles, Continental Oil Company, Exxon Corporation, Getty Oil Company, Gulf Oil Corporation, The Louisiana Land and Exploration Company, Marathon Oil Company, Mobil Oil Corporation, Murphy Oil Corporation, Petrofina Societe Anonyme, Phillips Petroleum Company, Royal Dutch/Shell Group of Companies, Skelly Oil Company, Standard Oil Company of California, Standard Oil Company (Indiana), The Standard Oil Company (Ohio), Sun Company, Inc., The Superior Oil Company, Texaco Inc., Tosco Corporation, and Union Oil Company of California.

4. For an explanation of variance analysis, see M. Ezekiel and K. Fox, *Methods of Correlation and Regression Analysis*, 3rd ed. (New York: John Wiley and Sons, 1959). The values of the F-ratio ranged from 0.1234 to 1.835. Detailed results will be furnished to any interested researcher on request.

5. See *PE*, October 1976. See Appendixes D to F for cost and investment comparisons.

6. In the Middle East, on the other hand, nationalization of foreign company assets has relegated most companies to providers of technology and services. As such they have been relieved from the 'burden' of providing 'risk capital' through all phases of the industry and are able to use their profits for diversification out of oil into other energy endeavours. See *AWSJ*, 19 October 1978, pp. 1, 9, 'International oil firms find OPEC still needs their technology, skill'.

7. See the *Oil and Gas Journal* year-end issues for production data by fields.

8. The term 'small' is used relatively and refers to the comparative size of the companies when viewed against the size of the 'Seven Sisters'. The seven international 'majors' often referred to as the 'Seven Sisters' are: Exxon (which was formerly Standard Oil Company, New Jersey), Royal Dutch/Shell Group, British Petroleum Company, Gulf Oil Corporation, Texaco, Standard Oil of California, and Mobil Oil Corporation.

9. On the other hand, over the long term there may be some serious questions on the effects of certain contract provisions—as a subsequent section points out.

10. See Chapter III again.

11. For a good illustration of this concept, see A. P. H. Van Meurs, *Petroleum Economics and Offshore Mining Legislation* (Amsterdam: Elsevier, 1971), pp. 120–2.

12. The notion of 'limbo oil' is another way of expressing this effect of the fiscal aspects of contracts. Stauffer refers to three categories of oil reserves: (1) windfall oil; (2) 'limbo oil'; and (3) non-economic oil. 'Limbo oil' refers to fields which could be produced economically but for which government 'take' leaves a margin too narrow for an entrepreneur to want to produce it. See Thomas R. Stauffer, 'The Economics of Petroleum Taxation in the Eastern Hemisphere', paper presented before the OPEC Seminar on 'International Oil and the Energy Policies of the Producing and Consuming Countries', Vienna, 30 June–5 July 1969.

13. A *contingent payment* is defined as that made only if a certain event (or set of events) comes to pass. Examples are royalties, profit shares or production shares. See H. E. Leland, R. B. Norgaard, and Scott Pearson, 'An Economic Analysis of Alternative Outer Continental Shelf Petroleum Leasing Policies', unpublished manuscript prepared for the Office of Energy R&D Policy, U.S. National Science Foundation, September 1974.

14. See Allen G. Hatley, 'Asia's Oil Prospects and Problems: An Overview of Petroleum Exploration Activity in East Asia', paper presented at the SEAPEX session, Offshore Southeast Asia Conference, 21–24 February 1978. Hatley presented an actual example where his company's management approved two recommended acquisitions whose contractual terms ranked economically higher but whose geological prospects ranked lower than those rejected (i.e., the technical risks were higher).

15. See the January issues of *Petroleum News* listing contractors.

16. M. A. Adelman, *The World Petroleum Market* (Baltimore: The Johns Hopkins University Press, for Resources for the Future, 1972), pp. 53–5.

17. L. K. Weaver *et al.*, *Composition of the Offshore U.S. Petroleum Industry and Estimated Costs of Producing Petroleum in the Gulf of Mexico* (Washington, D.C.: U.S. Department of the Interior, 1972), and E. W. Erickson and R. M. Spann, 'The U.S. Petroleum Industry', in E. Erickson and L. Waverman (eds.), *Energy Question: An International Failure of Policy* (Toronto: University of Toronto Press, 1974). See also Dennis E. Logue *et al.*, 'Optimal Leasing Policy for the Development of Outer Continental Shelf Hydrocarbon Resources', *Land Economics*, Vol. 51 (August 1975), pp. 191–207, and R. W. Bybee, 'Petroleum Exploration and Production on the Nation's Continental Shelves—Economic Potential and Risk', paper presented at the Annual Meeting of the Marine Technology Society, 1970.

18. Peter D. Gaffney and C. P. Moyes, 'Competitive Legislation—the Key to Asia Pacific Petroleum Prospects', paper presented at the Society of Petroleum Engineers' Session, Offshore Southeast Asia Conference, 21–24 February 1978, Figure 5.

19. *AWSJ*, 24 October 1978, p. 1, 'Thailand still faces hurdles in pricing Texas Pacific Gas'.

20. *OGJ*, 20 June 1977, p. 34, and Siddayao (1978), op. cit., pp. 124–57.

21. Adelman, op. cit., p. 53.

22. See Thomas Horst, 'American Taxation of Multinational Firms', *American Economic Review*, Vol. 67 (June 1977), pp. 376–89.

23. See Chapter III again.

24. See Hatley (1978), op. cit.

25. Interview in July 1978.

VI

Some Supply Variables and Patterns of Responses of Firms

IN this chapter, some of the more important exploration indicators in South-East Asia will be analysed in the context of the discussion in Chapter V. Cost and profit data are generally difficult to obtain, but it is even more difficult to try to estimate company costs in this region. The analyses will, therefore, focus on the technical-economic variables that are normally reported. Also, although the discussion will try to cover all of South-East Asia where possible, some of the specific examples will be drawn from Indonesia. The longer history of exploration and production in Indonesia and the relatively better accessibility of data make this approach practical.

THE APPROACH

The analysis below follows to a high degree the approach of Fisher in a very much earlier study.[1] Fisher's study has been superseded by other, more advanced models (such as those of Erickson, Erickson-Spann, MacAvoy-Pindyck, and Epple).[2] The paucity of appropriate data in South-East Asia, however, makes a partial test of the Fisher model a more useful exercise for our purposes.

Fisher's supply curve model was based on the following hypothesis: the wildcatting decisions made by the industry is a process whose inputs are various geophysical/geological and economic factors and whose outputs are the number of wildcats drilled and the average characteristics of the drilling results. Three major conclusions were suggested by Fisher: (1) wildcat drilling is highly sensitive to changes in prices; (2) the success ratio[3] is negatively related to price increases; (3) the discovery size is negatively related to price increases, and large discoveries tend to provide incentives for further wildcatting.

In the present study a stock model is used, with the following elements: The national government has a stock of petroleum reserves; it would like to maintain an inventory of these reserves at a level dictated by its foreign exchange requirements. The firm supplies the petroleum reserves via successful exploration wells, and the firm's decision to drill these wells depends on economic incentives. Underlying such incentives are the factors described in Chapter III, i.e., the institutional framework provided by the host government, the firm's international character, technological (cost) factors, world market prices, and the firm's utility-maximizing criteria, given its constraints.

The following variables were considered in this study: (1) the number of exploratory wells in South-East Asian countries; (2) the average depth of such wells; (3) the drilling cost index; (4) the success ratios of these exploration wells; (5) the size of discoveries; (6) the distribution of such discoveries; (7) world oil prices; and (8) institutional changes. The first three variables are cost-generating in nature; the next four variables are related to potential revenues from the venture; and the last may affect both actual and expected costs and revenues. Most of these data are presented in Tables 6.1 and 6.2. The price data were presented in Table 3.4.

In the following sections, a descriptive analysis is first undertaken. This descriptive analysis will be followed by the section presenting the results of statistical tests.

Qualitative Analyses

In this section, descriptive analyses of the following relationships are undertaken: (1) the relationship between wildcat success in the previous year to the level of exploratory drilling in the current year; (2) the relationship between world prices and drilling rates; and (3) the relationship between size of discovery and the level of exploratory drilling. For this purpose, data on South-East Asia are plotted in Figure 6.1. Data on Indonesia are plotted in Figure 6.2.

SUCCESS RATIOS AND DRILLING RATES

Table 6.1 shows comparative information on completed exploration wells in South-East Asia during the period 1967–76.[4] The Indonesian data provide a more interesting history than any of the countries, because of the longer and more prolific activity. There has also been relatively freer access to information on developments compared to those of Brunei, Burma, and Malaysia.

The table, however, may be somewhat misleading if one were to judge the prospects for petroleum development in each country only by the success ratios in the table. Vietnam, for example, had a success ratio of 33 per cent on the basis of two successful wells in six drilled in 1974 and 1975. This may be regarded as a high success ratio, but at the same time the level of activity may appear to indicate lack of interest. The level, as anyone knows who has followed political developments in the area, results from non-geologic factors. These were discussed earlier in Chapters II and III.

In the meantime, the Philippine discoveries off Palawan in 1977 and 1978 have upgraded the prospects of the Philippines as a potential petroleum region.[5] The seemingly sparse interest in Malaysia also belies the exploration successes of 1977 and 1978; the low level of activity in 1975 and 1976 coincided with the virtual standstill in exploration that resulted from a government shift in resource policies and the resulting renegotiation of contractual arrangements covering petroleum resource development. (See Chapter IV again.)

Figures 6.1 and 6.2 indicate the existence of some relationship between the total number of wells in the current year and the lagged values of price, discovery size and successful wells. It will be noted that the shapes of both the 'total' and the 'successful' curves for South-East Asia (Figure 6.1) approximate the shapes of the same curves for Indonesia (Figure 6.2). This follows from the high proportion of both drilled and successful wells accounted for by Indonesia.

Although a lagged direct relationship appears to exist between the two well variables—total and successful—the drop in total wells drilled in 1976 and 1977 does not appear to be justified by the success ratio. This will be taken up again in the discussion of the

TABLE 6.1
COMPLETED* EXPLORATION WELLS IN SOUTH-EAST ASIA, 1967–1977

Countries and Variables	1967	1968	1969	1970	1971	1972	1973	1974	1975	1976	1977
Brunei											
Total	12	4[a]	5	3	NA	6	9	NA	12	11	3
Successful	1	NA	2	1	NA	NA	NA	—	NA	7	2
Success ratio (%)	8	NA	40	33	NA	NA	NA	NA	NA	64	67
Depth (10³ m)	31.3	NA	15.0	NA	NA	NA	NA	NA	NA	31.1	10.3
(10³ ft)	(102.6)	NA	(49.3)	NA	NA	NA	NA	NA	NA	(102.1)	(33.9)
Average depth (10³ m)	2.6	NA	3	NA	NA	~1.7 (2)	~2.1 (3)	NA	~2.5[b]	2.8	3.4
(10³ ft)	(8.5)	NA	(9.9)	NA	NA	~(5.5)(2)	~(6.7)(3)	NA	~(8.2)	(9.3)	(11.3)
Burma											
Total	—	—	—	3	NA	7	11	6	>8	>10	NA
Successful	—	—	—	3	NA	1	1	—	0	0	NA
Success ratio (%)	—	—	—	100	NA	14	14	17	0	0	NA
Depth (10³ m)	—	—	—	>4.7[c]	NA	>10.9 (3)	>12.0 (4)	16.5	>26.7 (8)	>31.0	NA
(10³ ft)	—	—	—	(>15.5)	NA	(>35.7)(3)	(>39.3)(4)	(54.0)	(>87.7)(8)	(>101.7)	NA
Average depth (10³ m)	—	—	—	~1.6	NA	~3.6 (3)[d]	~3.0 (4)	2.7	~3.3 (8)	~3.1	NA
(10³ ft)	—	—	—	~(5.2)	NA	~(11.9)(3)[d]	~(9.8)(4)	9.0	~11.0 (8)	~10.2	NA
Indonesia											
Total	5	19	32	84	135[b]	137	169	167	183	126	103
Successful	0	5	6	17	12	34	30	46	49	38	33
Success ratio (%)	0	26	19	20	9	25	18	28	27	30	32
Depth (10³ m)	>3.9	13.5	57.9	>138.3	247.0	239.8	306.4	326.3	342.8	219.0	180.0
(10³ ft)	>(12.7)[c]	(44.4)	(190.0)	>(453.5)	(810.3)	(786.9)	(1,004.6)	(1,069.8)	(1,124.6)	(718.4)	(592.2)

Countries and Variables	1967	1968	1969	1970	1971	1972	1973	1974	1975	1976	1977
Indonesia (continued)											
Average depth											
(10³ m)	0.8	0.7	1.8	~1.8 (76)e	2.1	1.7	1.8	2.0	1.9	1.7	1.8
(10³ ft.)	(2.6)	(2.3)	(5.9)	~(6.0)(76)e	(7.0)	(5.7)	(5.9)	(6.4)	(6.2)	(5.7)	(5.8)
Kampuchea (Khmer)											
Total	—	—	—	—	—	1	—	2	—	NA	NA
Successful	—	—	—	—	—	0	—	0	—	NA	NA
Success ratio (%)	—	—	—	—	—	0	—	0	—	NA	NA
Depth											
(10³ m)	—	—	—	—	—	2.4	—	3.7	—	NA	NA
(10³ ft.)	—	—	—	—	—	(8.0)	—	(12.1)	—	NA	NA
Average depth											
(10³ m)	—	—	—	—	—	2.4	—	1.8	—	NA	NA
(10³ ft.)	—	—	—	—	—	(8.0)	—	(6.1)	—	NA	NA
Malaysia											
Total	13	13	22	44	32	32	26	26	40	8	11
Successful	NA	NA	<6>f	<6>f	<6>f	2	3	4	3g	0	1
Success ratio (%)	NA	NA	~27	~14	~19	6	12	15	8	0	8
Depth											
(10³ m)	29.6	37.6	43.9	79.1	68.8	70.3	55.4	59.0	90.3	18.8	23.2
(10³ ft.)	(97.0)	(123.3)	(144.2)	(259.6)	(225.7)	(230.5)	(181.9)	(193.7)	(296.2)	(61.7)	(76.0)
Average depth											
(10³ m)	2.3	2.9	2.0	2.3	2.2	2.2	(2.1)	2.3	2.3	2.4	2.1
(10³ ft.)	(7.5)	(9.5)	(6.6)	(7.6)	(7.0)	(7.2)	(7.0)	(7.4)	(7.4)	(7.7)	(6.9)
Philippines											
Total	—	—	—	2	17	8	11	7	12	8	12
Successful	—	—	—	0	0	0	0	0	0	0	2
Success ratio (%)	—	—	—	0	0	0	0	0	0	0	17

TABLE 6.1 (continued)

Countries and Variables	1967	1968	1969	1970	1971	1972	1973	1974	1975	1976	1977
Philippines (continued)											
Depth											
(10³ m)	—	—	—	1.1	27.5	14.3	21.0	17.9	30.7	21.7	32.6
(10³ ft.)	—	—	—	(3.6)	(90.3)	(47.1)	(69.0)	(58.6)	(100.6)	(71.2)	(106.9)
Average depth											
(10³ m)	—	—	—	.6	1.6	1.8	1.9	2.6	2.6	2.6	2.7
(10³ ft.)	—	—	—	1.8	5.3	(5.9)	(6.3)	(8.4)	(8.4)	(8.4)	(8.9)
Thailand											
Total	—	—	—	—	1	5	6	14	10	17	2
Successful	—	—	—	—	0	0	2	3	2	1	2
Success ratio (%)	—	—	—	—	0	0	33	21	20	6	100
Depth											
(10³ m)	—	—	—	—	2.9	8.7	14.1	33.6	26.5	46.7	5.3
(10³ ft.)	—	—	—	—	(9.6)	(28.5)	(46.1)	(110.0)	86.8	(153.3)	(17.3)
Average depth											
(10³ m)	—	—	—	—	2.9	1.7	2.4	2.4	2.6	2.8	2.6
(10³ ft.)	—	—	—	—	(9.6)	(5.7)	(7.7)	(7.9)	8.7	(9.0)	(8.6)
Vietnam											
Total	—	—	—	—	—	—	—	2	4	—	NA
Successful	—	—	—	—	—	—	—	1	1	—	NA
Success ratio (%)	—	—	—	—	—	—	—	50	25	—	NA
Depth											
(10³ m)	—	—	—	—	—	—	—	5.7	11.8	—	NA
(10³ ft.)	—	—	—	—	—	—	—	(18.7)	(38.6)	—	NA
Average depth											
(10³ m)	—	—	—	—	—	—	—	2.8	2.9	—	NA
(10³ ft.)	—	—	—	—	—	—	—	(9.3)	(9.7)	—	NA

Countries and variables	1967	1968	1969	1970	1971	1972	1973	1974	1975	1976	1977
South-East Asia											
Total	30	36	59	>136	>185	196	232	>224	269	>180ᶜ	131
Successful	1	5ᶜ	14ᶜ	27ᶜ	18ᶜ	37ᶜ	36ᶜ	55ᶜ	55ᶜ	46ᶜ	40ᶜ
Success ratio (%)	6(17)	26(19)	24	20ᶜ	10(18)ᵇ	20ᶜ	16ᶜ	25ᶜ	20ᶜ	26(180)ᶜ	30(131)ᶜ
Depthᶜ											
(10³ m)	64.7	>51.1(32)	116.9	>223.2(125)	>346.2(185)	>349.8(183)	>414.9(219)	>462.4(224)	>528.7(257)	>367.1	251.9
(10³ ft.)	(212.3)	>(167.7)	(383.5)	>(732.2)	>(1,135.9)	>(1,147.7)	>(1,361.1)	>(1,516.9)	>(1,734.5)	>(1,204.4)	(826.3)
Average depthᶜ											
(10³ m)	2.2	~1.61(32)	2.0ᵈ	~1.4(125)	~1.9(172)ᵇ	~1.9(183)	~1.9	~2.1	~2.1	~2.0(180)	1.9(131)
(10³ ft.)	(7.7)	~(5.2)	(6.5)	~(4.6)	~(6.1)	~(6.3)	~(6.2)	~(6.8)	~6.7	~6.7	(6.3)

Sources: Based on data in the American Association of Petroleum Geologists, *Bulletin* (both text and tables of the 'Foreign Developments' issue); in the *International Petroleum Encyclopedia*, the *Oil and Gas Journal* (26 December 1977 and 25 December 1978), and UNDP/CCOP, *Technical Bulletin No. 11*.

Note: Numbers in parentheses next to depth and success ratio figures indicate number of wells for which data were available.

*Only completed wells are listed for the year. Wells drilled in one year and completed in the next will be listed in the year of completion. Totals may not coincide with those in the *International Petroleum Encyclopedia;* this table relies heavily on the AAPG *Bulletin,* both text and tables, and contains *IPE* data only where data are absent in the *Bulletin.* It also contains data on Malaysia released by the Malaysian government to and published by the UNDP/CCOP.

ᵃ Breakdown of figures not available.

ᵇ Based on data for 38 extension and development wells. See AAPG (1977).

ᶜ Data incomplete.

ᵈ Offshore only. No onshore data available, but absence of exploratory activity not verifiable.

ᵉ Data for 76 wells only. Caltex drilled 8 wells, with 3 successful; but no depth data available.

ᶠ Sarawak Shell discovered 18 wells (12 oil and 6 gas) in 1969, 1970, 1971 (AAPG, 1972).

ᵍ Estimated. Results were not released by Exxon for 3 wells drilled in Tapis field, later reported as discovered in 1975.

TABLE 6.2

DATA COMPARISON OF COMMERCIAL OIL FIELDS IN SOUTH-EAST ASIA DISCOVERED 1967-1976*

Countries and Variables	1967	1968	1969	1970	1971	1972	1973	1974	1975	1976
Brunei										
No. of fields	1	0	1	1	0	0	0	0	0	0
(No. offshore)	(1)	—	(1)	(1)	—	—	—	—	—	—
Cumulative production to 1 July 1978 (10³ bbl.)	6,330	—	57,206	92,234	—	—	—	—	—	—
Average (approximate)[a] annual output per field (10³ bbl.)	633	—	7,150	13,176	—	—	—	—	—	—
Range of peak daily production during 1974-8 (mbd)	7,122	—	29,586	76,564	—	—	—	—	—	—
Average depth (ft.)	9,500	—	10,740	4,300	—	—	—	—	—	—
Burma										
No. of fields	0	0	0	1	0	0	0	0	0	0
(No. offshore)	—	—	—	(0)	—	—	—	—	—	—
Cumulative production to 1 July 1978 (10³ bbl.)	—	—	—	30,050[b]	—	—	—	—	—	—
Average (approximate)[a] annual output per field (10³ bbl.)	—	—	—	4,292	—	—	—	—	—	—
Range of peak daily production during 1974-8 (mbd)	—	—	—	17,300	—	—	—	—	—	—
Average depth (ft.)	—	—	—	4,300	—	—	—	—	—	—

Countries and Variables	1967	1968	1969	1970	1971	1972	1973	1974	1975	1976
Indonesia										
No. of fields	0	6	4	7	11	15	18	12	8	6
(No. offshore)	(0)	(0)	(1)	(2)	(0)	(5)	(2)	(3)	(0)	(1)
No. COW/PSC[c]	—	6	3	3	8	5	10	5	6	2
No. PSC[d]	—	0	1	2	3	9	7	4	2	4
(PSC offshore)	—	—	(1)	(2)	(0)	(5)	(2)	(3)	(0)	(1)
Cumulative production to 1 July 1978 (10^3 bbl.)	—	88,718	256,672	576,195	137,656	137,167	139,469	204,563	71,611	22,655
Average (approximate)[a] annual output per field (10^3 bbl.)	—	1,643	8,021	11,759	2,085	1,828	1,937	5,682	4,476	3,778
Range of peak daily production during 1974–8 (mbd)	—	1,146–21,252	2,962–113,416	3,506–112,505	627–42,967	16–55,718	781–24,444	526–171,718	194–46,553	1,258–46,880
Average depth (ft.)	—	4,530	3,747	4,277	3,944	4,298	4,453	4,849	3,266	3,993
Malaysia										
No. of fields	1	0	0	0	2	1	2	0	1	0
(No. offshore)	(1)	—	—	—	(2)	(1)	(2)	—	(1)	—
Cumulative production to 1 July 1978 (10^3 bbl.)	59,880	—	—	—	13,868	54,594	21,503	—	1,731	—
Average (approximate)[a] annual output per field (10^3 bbl.)	5,988	—	—	—	1,156	10,918	2,688	—	865	—
Range of peak daily production during 1974–8 (mbd)	52,900	—	—	—	3,144–12,930	69,716	10,236–27,096	—	9,565	—
Average depth (ft.)	11,000	—	—	—	13,500	8,000	5,650	—	6,700	—

TABLE 6.2 (continued)

Countries and Variables	1967	1968	1969	1970	1971	1972	1973	1974	1975	1976
South-East Asia (Summary)										
No. of fields	2	6	5	9	13	16	20	12	9	6
(No. offshore)	(2)	(0)	(2)	(3)	(2)	(6)	(4)	(3)	(1)	(1)
Cumulative production to 1 July 1978 (10^3 bbl.)	66,210	88,718	313,878	698,479	151,524	191,761	160,972	204,563	73,342	22,655
Average (approximate)[a] annual production per field (10^3 bbl.)	3,310	1,643	7,847	11,087	1,943	2,397	2,012	5,682	4,075	3,777
Range of peak daily production	7,122–	1,146–	2,962–	3,506–	627–	16–	781–	526–	194–	1,258–
	52,900	21,252	113,416	112,505	42,967	69,716	27,096	171,718	46,553	46,880
Average depth (ft.)	10,250	4,530	5,145	4,582	5,414	4,529	4,572	4,849	3,825	3,993
(Average offshore)	(10,250)	—	(8,119)	(5,100)	(13,500)	(6,292)	(6,288)	(5,514)	(6,700)	(2,139)

Source: Based on data in year-end issues of the *Oil and Gas Journal* to 1978.

*No fields discovered in 1977 were reported to be in commercial production as of the end of 1978.

[a] Estimated as $(Q/n\text{-}2) \div F$, where Q = cumulative output; n = no. of years, including year of discovery and year of reporting (1 July 1978); F = no. of fields.

[b] Estimated.

[c] COW/PSC = Contract-of-work/Production-sharing contract holders, namely, Caltex and Stanvac.

[d] PSC = production sharing.

Note: Balance of discovered fields are in government developed areas.

results of statistical tests on the different variables. At this point, however, it might be useful to note the effects of the fall of the South Vietnam government in 1975, and the contractual and other institutional changes in Indonesia and Malaysia of 1976.

DRILLING RATES AND PRICE CHANGES

From the plot of prices taken from Table 2.4 and the trend line for total wildcats in Figure 6.1, we note a general positive relationship between changes in international oil prices and drilling. It would be useful to recall at this point, however, that the increasing trend in drilling also reflected the effects of contract requirements. (See Chapter IV again.) By 1971 the bulk of PSC holders in Indonesia were in the second or third years of their exploration period. (The effect of price on drilling will be taken up again in the section on statistical results.)

DISCOVERY SIZE AND DRILLING RATES

Table 6.2 shows a summary of fields discovered in South-East Asian countries during the years 1967 to 1976, the size range (lowest to highest) of the fields discovered as indicated by the highest daily output during the years 1974 to 1978, and the average (estimated) output per field. The table separates Indonesian discoveries in PSC areas from those in the areas of former concession holders, who concurrently hold contracts of work and production sharing contracts.

The data on average size of discoveries are also plotted in Figures 6.1 and 6.2, to see the relationship between size of discoveries in year n to the exploratory drilling trend in year $n+1$. A general direct relationship between size of discovery and exploratory drilling levels appears to exist. Again, the drop in exploratory drilling in 1975 and 1976 following the size of discoveries in 1975 does not follow intuitive reasoning and obviously reflects once again non-economic factors, perhaps the reaction of firms to the 1976 contract revisions in Indonesia. It might be interesting to note also the sharp drop in the average size of the 1976 discovery. Whether this represented the results of the contractor's 'best efforts' may, of course, be questioned, from the viewpoint of those who are inclined

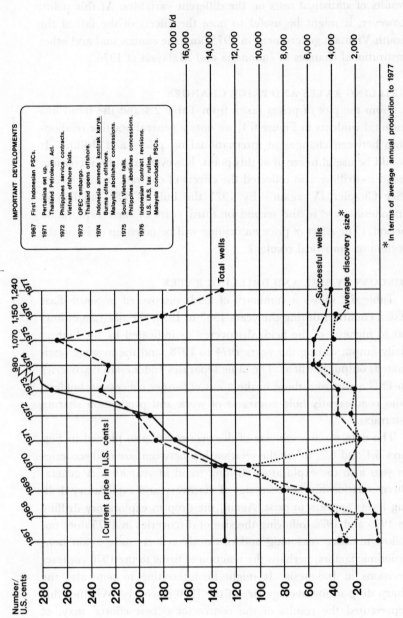

Figure 6.1. Exploration in South-East Asia

IMPORTANT DEVELOPMENTS

1967 First Indonesian PSCs.
1971 Pertamina set up.
      Thailand Petroleum Act.
1972 Philippines service contracts.
      Vietnam offshore bids.
1973 OPEC embargo.
      Thailand opens offshore.
1974 Indonesia amends kontrak karya.
      Burma offers offshore.
      Malaysia abolishes concessions.
1975 South Vietnam falls.
      Philippines abolishes concessions.
1976 Indonesian contract revisions.
      U.S. I.R.S. tax ruling.
      Malaysia concludes PSCs.

Total wells

Successful wells

Average discovery size*

* In terms of average annual production to 1977

(Current price in U.S. cents)

Number/
U.S. cents

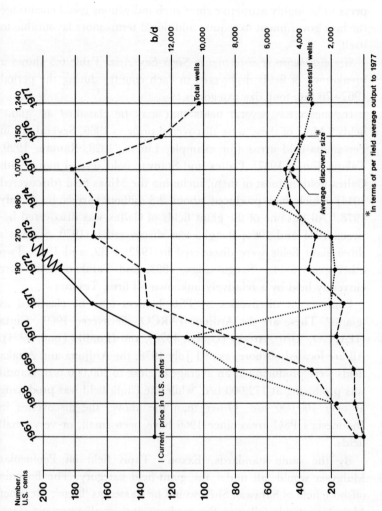

Figure 6.2. Exploration in Indonesia

* In terms of per field average output to 1977

to suspect companies of incomplete reporting. The contract revisions certainly provided no incentive for the companies to provide the government with more grounds for judging the Indonesian prospects to be highly attractive since such indications could encourage the host government to require additional terms more favourable to itself.

Size distribution of discoveries in South-East Asia. Table 6.3 shows a breakdown of fields discovered in each country during the period 1967–76 into four size categories.

Indonesia has several fields that may be classified as 'giant' fields.[6] Most of these were discovered prior to 1966. Several are in Pertamina-held areas (for example, Limau, 1928, Rantau, 1929, Talang Jimar, 1937). Caltex and Stanvac hold several more, with Caltex holding most of them, including the Minas field (discovered 1944) which had produced about 2.3 billion barrels by 1 July 1978.[7] All but one of the giant fields of Caltex was discovered between 1941 and 1964; Bangko was discovered in 1970. Stanvac's three giant fields were discovered in 1922, 1932, and 1940. Two other giant fields—Sanga-sanga, 1893, and Tarakan, 1906—are currently held by a relatively unknown oil firm, Tesoro.

Four recent discoveries by PSC holders may be classified as 'giants'. These are the Ardjuna (ARCO, discovered 1969), Cinta (IIAPCO, 1970), Attaka (Union, 1970), and Handil (Total, 1974). All are located offshore; as of 1 July 1978, the Ardjuna and Attaka fields were producing at an average of close to 200,000 b/d; Handil was producing at 172,000 b/d, while the Cinta field was producing at over 100,000 b/d. Other than the above, the discoveries in Indonesia's PSC areas since 1966 have been small, or very small fields.

By the same standards, Exxon's Tapis field off Peninsular Malaysia would fall under the giant-field category. The Baronia offshore field of Sarawak Shell would be classed as 'large'; the other Malaysian fields fall into the medium and small reservoir class. (See Table 6.3 again.)

The Philippine discoveries (Nido and Matinloc) off Palawan in 1977 and 1978 are reported to have recoverable oil totalling 100 million barrels;[8] at this writing, however, the sizes of the fields have

TABLE 6.3
SIZE DISTRIBUTION OF COMMERCIAL FIELDS
DISCOVERED 1967–1976

Country	Year	Giant	Large	Medium	Small	Total
Brunei	1967	0	0	0	1	1
	1969	0	0	1	0	1
	1970	0	1	0	0	1
Burma	1970	0	0	1	0	1
Indonesia	1967	0	0	0	3	3
	1968	0	0	1	5	6
	1969	1	0	1	2	4
	1970	3	1	1	2	7
	1971	0	0	2	9	11
	1972	0	1	1	13	15
	1973	0	0	5	13	18
	1974	1	1	2	8	12
	1975	0	1	0	7	8
	1976	0	1	0	5	6
Malaysia	1967	0	1	0	0	1
	1971	0	0	0	2	2
	1972	1	0	0	0	1
	1973	0	0	1	1	2
	1975	0	0	0	1	1
Total South-East Asia		6	7	16	72	101

Source: Based on data in year-end issues of the Oil and Gas Journal to 1978. Definitions based on National Petroleum Council, Offshore Petroleum Resources (1975), Table 9.

Definitions:

Giant	= >100,000 b/d	Medium	=	>15,000 b/d
Large	= >50,000 b/d	Small	=	± 15,000 b/d

not been adequately assessed, although there were reports that a platform with 50,000 b/d capacity was planned for the Nido field.[9] Reports on its actual size have been conflicting. It was indicated in early 1979 that the Nido field was expected to produce at 40,000 b/d[10]—which would put it in the medium-to-large size category. Thailand's gas discoveries fall under the 'major strike' classification of the AAPG (see Chapter II again).

Distribution of discoveries in Indonesia. How are the discoveries in Indonesia distributed between the two groups of contract holders? As Table 6.2 shows, the contract-of-work holders consisting basic-ally of Caltex and Stanvac (plus their affiliates) accounted for about 55 per cent of the fields discovered between 1967 and 1976. All of these discoveries were onshore. Perhaps this was to be expected, since Caltex holds most of the more prolific onshore areas.

On the other hand, in terms of size, the PSC holders discovered four or five 'giant' fields during this period. In addition, about 40 per cent of the fields discovered by the PSC holders during the period under review were in offshore areas.

Interestingly, most of the largest finds for both groups occurred during the years 1969–70. It has often been said that the Indonesian oil fields that remain to be discovered are small in size, so that continued exploration at a higher degree than in other re-gions was necessary to locate new fields and keep the country's reserves from petering out to precariously low levels. One industry geologist indicated that most Indonesian fields recently discovered have production lives of as short as three years, some with five, and others with ten. The small fields become commercially feasible to develop when they are located close enough to each other to allow consolidation of investments in production facilities.

The effect of the size of the discovery on exploratory drilling will again be taken up in the following section on quantitative esti-mates.

Quantitative Inferences

In this section, some of the results of several attempts to apply statistical estimation methods on the importance of certain supply variables will be presented.

The exploration data are mainly those presented in Tables 6.1, 6.2, and 6.3. The price data are from the World Bank, and the cost data are from the U.S Bureau of Labor Statistics. We realize that our exploration data are less than complete, and are not in the desired detail or form. As a result, the product of the efforts at estimation has been far from satisfactory.

It is not, in fact, customary to report results that are only partially successful. The present study is, however, a pioneering work on South-East Asia's petroleum industry, and we thought doing so might be useful to others who may wish to inquire further into this problem. We do not pretend that the quantitative estimates presented here may be used with any confidence, or that we can generalize on the basis of the results. At the same time, however, the results provide some insight into the relative importance of certain variables in the supply of petroleum reserves in South-East Asia.

THE SUPPLY VARIABLES ESTIMATED

In this part of the study, we made the following assumptions. Since petroleum reserves can only be supplied by drilling, the number of exploratory wells (or wildcat wells, often simply referred to as 'wildcats') may be used as an indicator of the petroleum firm's decision to supply such reserves. This decision is subject to several constraints, and is influenced by several factors, each of which will be discussed briefly below.

The functional relationship between the number of wildcat wells drilled and the variables considered in this study may be expressed as follows:

$$W_{jt} = S(S_{jt-1}, P_{t-1}, C_{t-1}, R_{jt-1}, S^i_{jt-1}, D_{t-1}, DU_{jt-1}, DU_{jt})$$

where W = wildcat wells drilled, P = world oil prices (expressed in constant values), C = drilling costs, R = the success ratio, S = the size of oil discovery per field, D = the average depth per well, DU = a dummy variable to reflect institutional change, t = the current year, j = the specific country, and i = 1, 2, 3, 4 (to distinguish the four categories of field sizes).

The inclusion of the lagged variable for wildcat wells was based on the following observation and assumptions. Contracts require some minimum drilling obligation shortly after an award. After

completion of this requirement, further drilling on the same contract area will depend highly on the results of the first wells, hence
the inclusion of the variables on the success ratio and the size of
discovery. New drilling will depend on new contract awards.

Since most of the countries were either not producing or were
producing in a limited way at the time the exploration wells were
drilled, and also since the companies involved are generally international companies, the world market price data were considered
to be relevant. Even in the case of Indonesia, which produces
mainly for export, this appears relevant. Absence of cost data on
the area, both as a result of host country and company restrictions
as well as of the relatively unknown nature of the area, made the
use of the U.S. cost index on drilling machinery, equipment, and
supplies acceptable. The fact that drilling expertise and capital
equipment are mainly of U.S. origin also made the use of the U.S.
cost index acceptable.

Theory tells us that a price increase should influence wildcat
drilling in a direct relationship, that is, a substantial price increase
should serve as an incentive to exploration drilling in higher cost
areas, assuming that footage costs remain unchanged. Hence, we
expect the price variable to have a positive sign. On the other
hand, theory suggests that cost increases would discourage drilling
if prices did not rise commensurately. In this case, we expect the
cost variable to have a negative sign. The depth variable actually
serves as an indicator of costs in the region, and so we expect it to
have a negative sign as well. However, we expect a possible correlation between the two variables.

The expected signs of the variables on discoveries are unclear
and depend on how one would like to interpret the firm's expectations. Fisher obtained negative signs on the success ratio variable,
and explained this result in this manner: 'A rise in the success ratio
in the past year does not necessarily mean a rise in the ease of
finding oil on prospects considered last year or this year; rather, it
means an acceptance of relatively certain prospects last year, and
this is by no means the same thing.' We find this explanation
rather unsatisfactory. At any rate, one might expect a high success
ratio to have a positive effect on exploratory drilling in South-East

Asia, because of the frontier nature of the region.

This direct relationship may be expected of the size-discovery variable. However, we probably should expect that the firms know that there is a distribution of sizes, so that a large discovery in one year dampens expectations of a similar discovery the following year. On the other hand, firms, being so aware, may be prepared to discover a small, but 'acceptable' size of field.

Inclusion of the average depth variable was thought desirable as an indicator of both the cost and the risks involved in exploration. At the same time, although the success ratio and the discovery size variables may be viewed as indicators of the profitability of the project, these variables may also be used as indicators of the probability of investment loss.

Use of the dummy variable to represent the presence or absence of any change in any given year in the institutional environment was considered necessary in the context of South-East Asia. The changes are those described in Chapters III and IV.[11] It was initially assumed, however, for purposes of this estimation, that the change had a one-time effect; i.e., whether the effect was positive or negative, this change would cause an adjustment in the firm's drilling decision variables the year after the change but that the firm would proceed on the basis of this adjustment. It is probably more realistic to assume that such change actually has a distributed lag effect, and so an additional run was attempted, which assumed that any institutional change had an effect on drilling decisions in the year it occurred as well as in the following year. One of the results is given in Equation I(D). Distribution over a longer time period was not attempted because of the data constraint.

The significant number of gaps in data on Brunei precluded use of the wildcat information from that country. Data on Cambodia were also excluded from the estimation because only two observations were available.

INTERPRETATION OF THE RESULTS

Table 6.4 presents a comparison of the estimation results from using two methods as well as by the inclusion and exclusion of certain variables. The parameters were estimated using normal

TABLE 6.4
WILDCAT SUPPLY EQUATION COEFFICIENTS[a] (SELECTED EXPLANATORY VARIABLES EXCLUDED)

PARAMETER ESTIMATES

Independent Variables (Lagged)	Eq. I(A) (OLS)[b]	Eq. I(B) (Auto)[c]	Eq. I(C) (Auto)[c]	Eq. I(D) (Auto)[c]	Eq. II(A) (OLS)[b]	Eq. II(B) (Auto)[c]	Eq. II(C) (Auto)[f]
Wildcat	0.50 (4.55)***	0.52 (5.50)*	0.53 (5.80)*	0.54 (5.74)*	N —	N —	N —
Price	1.07 (0.81)	1.09 (0.84)	1.27 (1.00)****	0.89 (0.68)	0.67 (1.07)***	0.82 (1.33)***	0.56 (0.96)
Cost	−0.17 (−1.30)***	−0.19 (−1.44)***	−0.23 (−0.30)	−0.15 (−1.11)****	N —	N —	N —
Success ratio	0.16 (0.91)	0.13 (0.77)	0.97 (0.60)	0.12 (0.71)	N —	N —	N —
Giant discovery	24.30 (6.15)*	23.90 (6.97)*	23.98 (7.21)*	24.22 (7.14)*	29.04 (5.92)*	29.38 (5.76)*	26.6 (8.9)*
Large discovery	4.01 (0.55)	4.34 (0.70)	3.19 (0.52)	4.89 (0.79)	22.61 (2.99)**	19.1 (2.30)**	
Medium discovery	0.49 (0.12)	2.19 (0.54)	1.87 (0.46)	3.16 (0.77)	7.92 (1.71)**	6.71 (1.51)****	9.6 (17.5)*
Small discovery	5.96 (4.39)*	5.43 (4.57)*	5.59 (4.80)*	5.06 (4.21)*	10.1 (7.30)*	10.42 (7.42)*	

PARAMETER ESTIMATES

Independent Variables (Lagged)	Eq. I(A) (OLS[b])	Eq. I(B) (Auto[c])	Eq. I(C) (Auto[c])	Eq. I(D) (Auto[c])	Eq. II(A) (OLS[b])	Eq. II(B) (Auto[c])	Eq. II(C) (Auto[c])
Average depth	N —	N —	0.86 (1.47)***	N —	N —	N —	N —
Dummy[d]	6.67 (1.42)***	3.39 (0.70)	3.29 (0.69)	2.32 (0.48)	8.58 (1.54)***	9.82 (1.89)***	8.18 (1.59)***
Dummy (not lagged)	N —	N —	—	−6.66 (−1.56)***	N —	N —	N —
Constant	21.74 (1.68)**	24.71 (1.96)**	23.86 (1.93)**	21.41 (1.68)**	4.35 (1.04)***	3.38 (0.73)	4.58 (1.08)***
R^2 (D.F.)	0.97(32)	0.98(32)	0.98(31)	0.98(31)	0.94(35)	0.92(35)	0.93(37)
F-statistic (D.F.)	101.38* (9, 32)	N —	N —	N —	89.19* (6, 35)	N —	N —

[a] The figures in parentheses below each coefficient are the absolute values of the t-statistics.
[b] OLS = Ordinary least squares method.
[c] Least squares estimation by Cochrane-Orcutt type procedure (convergence = 0.001).
[d] Dummy for institutional change.

N = Not used.
D.F. = Degrees of freedom.
*Significant at 0.005 level.
**Significant at 0.05 level.
***Significant at 0.10 level.
****Significant at 0.20 level.

regression methods.[12] This included an iterative method to correct for autoregressive disturbances.[13] The equations were all linear.[14] The results of using two methods and of excluding certain variables are included because not one of the approaches yielded statistically significant results simultaneously for all variables.[15] An analysis of the various results, however, provides some hints concerning the importance of the variables studied. As Kmenta points out:

...it is quite possible to find that *none* of β_2, β_3, ..., β_k is significantly different from zero according to the t test and, at the same time, to reject the hypothesis that $\beta_2 = \beta_3 = ... = \beta_k = 0$ by the F-test. This could arise in the case where the explanatory variables are highly correlated with each other. In such a situation the separate influences of each of the explanatory variables on the dependent variable may be weak while their joint influence may be quite strong.[16]

It will be noted that the coefficient of determination (R^2)[17] for the equations I(A) to I(C) are 0.97 and up, with the R^2 value increasing with correction for autoregressive disturbances in Equations I(B) and I(C). Although the R^2s are lower for Equations II(A) and II(B), they still indicate that over 90 per cent of the variations in wildcat drilling may be explained by the price variable and the variables on discovery size and institutional change.

Unfortunately, however, not all the coefficients of the variables are statistically significant.

The wildcat lagged variable has coefficients which are consistently statistically significant (see Equations I(A) to I(C)).[18] This is in agreement with the assumption that drilling in the previous year influences current drilling rates. The contract requirement for drilling is only a partial explanation for this influence, however; a drop in current drilling also implies that no new contracts were awarded in the previous year that required drilling in the current year. When the wildcat variable was dropped along with other variables, the explanatory power of the equation only dropped by a few percentage points, and we are left with the impression that it may only be important to a small degree.

The price variable failed to show a statistically significant influence on wildcat drilling. Whether this resulted from the limited data or the failure to account for the influence of some unspecified but important variable is not clear. Although not statistically significant, however, the coefficients provide a hint, by virtue of their positive sign, of the expected direct relationship between price and the number of wildcats drilled.

The expected possible multicollinearity between the success-ratio variable and the discovery-size variables was evident in the results.[19] Exclusion of the former variable in the No. II equations improved the t-values for the discovery-size variables. Moreover, the coefficients had the expected signs.

The coefficient for the cost variable had the correct sign but it was not statistically significant. The expected correlation of this variable with the depth variable is evident from Equation I(C), where the addition of the depth variable caused the cost variable to be 'insignificantly different from zero'.

The estimated effects of the four discovery-size variables are unsatisfactory. Examination of the simple correlation matrix showed some strong correlation between the 'large' and 'giant' discovery variables and the 'medium' and 'small' discovery variables. What this probably tells us is that the firms are somewhat indifferent between sizes at the lower end of the size ladder and the sizes at the upper end. Additional estimations were, therefore, made, collapsing the four variables into two (that is, giant-large and medium-small); the coefficients for these two new variables were statistically significant.[20] (See Equation II(C).)

The last of the explanatory variables used is the dummy variable on institutional change. Although the coefficient was statistically significant only in Equation II(C), it had the correct sign, i.e., it showed a direct relationship between an institutional change and the drilling level. What Equations II(A), II(B), and II(C) tell us is that the size of the discoveries (which may be viewed as expectations of returns on investment) and probably institutional change (which affects such expectations) explain most of the variations in wildcat drilling.

Although the results of the quantitative tests are mixed, it appears possible to conclude that economic factors (returns on investments) and events that would influence such factors do influence the level of exploratory drilling and therefore the rate at which petroleum reserves will be supplied in South-East Asia. Only the size of the impact of each variable is unclear.

1. Fisher, *Supply and Costs in the U.S. Petroleum Industry: Two Econometric Studies.*

2. See E. W. Erickson, 'Economic Incentives, Industrial Structure and the Supply of Crude Oil Discoveries in the U.S., 1946–58/59' (unpublished manuscript); E. W. Erickson and R. M. Spann, 'Supply Response in a Regulated Industry: the Case of Natural Gas', *Bell Journal of Economics and Management Science*, Vol. 2, No. 1 (Spring 1971), pp. 94–121; P. W. MacAvoy and R. S. Pindyck, 'Alternative Regulatory Policies for Dealing with the Natural Gas Shortage', *Bell Journal of Economics and Management Science*, Vol. 4, No. 2 (Autumn 1973), pp. 454–98; and D. Epple, *Petroleum Discoveries and Government Policy* (Cambridge, Mass.: Ballinger Publishing Co., 1975).

3. The success ratio is the proportion of successful wells to total wells drilled.

4. The data appearing in the tables may differ from those reported by the host governments in official reports as well as in publications of the CCOP. For one thing, the success ratios for exploration wells are consistently lower. The disparities arise from differences in the method of reporting 'exploration wells'. Certain wells on the borderline between exploration and development, which are considered by industry as 'development wells', are reportedly classified by some governments as 'exploration wells'. To allow comparison with data in other countries and regions, we have elected to use data mainly appearing in the *Bulletin* of the American Association of Petroleum Geologists on the assumption that the categories follow standard industry practice. The *Oil and Gas Journal* and the *International Petroleum Encyclopedia* are·used only to supplement the AAPG source.

5. See, for example, Allen G. Hatley, 'The Nido Reef Oil Discovery in the Philippines—Its Significance', paper presented at the ASEAN Council on Petroleum (ASCOPE) Conference, 11–13 October 1977, Jakarta, Indonesia. See also *AWSJ*, 8 December 1978, pp. 1, 11, 'Major oil find considered likely for Philippines'.

6. The field size classifications were discussed in Chapter II.

7. *OGJ*, 25 December 1978, p. 120.

8. *OGJ*, 25 December 1978. See also *Asia Research Bulletin*, 31 January 1979.

9. *AWSJ*, 5 May 1978, p. 3.

10. *AWSJ*, 23 March 1979, p. 3, 'Philippines expects oil field to yield 40,000 barrels daily'.

11. Institutional change includes factors internal and external to the region. They include those factors listed in Figure 6.1.

12. Those who are unfamiliar with regression techniques may wish to consult J. Kmenta, *Elements of Econometrics* (New York: The Macmillan Company, 1971), and

M. Ezekiel and K. Fox, *Methods of Correlation and Regression Analysis*, 3rd ed. (New York: John Wiley, 1959).

13. See Kmenta, op. cit., Chapter 8. It is possible that the disturbance occurring at one point of observation is correlated with a previous disturbance, e.g., that the effect of a disturbance occurring in one period carries over to another period. If this happens, the results obtained with the ordinary least squares method are biased, and some form of correction becomes necessary. Least squares estimation by the Cochrane-Orcutt type procedure (convergence = 0.001) was used. (See D. Cochrane and G. H. Orcutt, 'Application of Least Squares Regression to Relationships Containing Autocorrelated Error Terms', *Journal of the American Statistical Association*, Vol. 44 (1949), pp. 32–61.) Ridge regression was employed to reduce variance but did not improve the results. (See A. E. Hoerl and R. W. Kennard, 'Ridge Regression: Biased Estimation for Nonorthogonal Problems', and 'Ridge Regression: Applications to Nonorthogonal Problems', *Technometrics*, February 1970, pp. 55–82.)

14. Log-linear functions were attempted with price and discovery size as independent variables, but the results were totally unsatisfactory.

15. The term 'statistical significance' cannot be satisfactorily defined in a brief manner. Interested readers may refer to any textbook on statistics; see also Kmenta, op. cit., Chapter V.

16. See ibid., p. 367.

17. The *coefficient of determination* may be said to measure the percentage of the variation in the dependent variable (in this case, the wildcats in year t) determined by the explanatory variables.

18. These were significant at the 0.005 level.

19. This was especially clear in the simple correlation matrix.

20. Statistically significant at the 0.005 level.

VII

Conclusions and Implications

MOST of the conclusions reached in this study were already suggested in the discussions. In this final chapter these conclusions will be crystallized and linked to major policy implications—the goal of this study.

The study's objective was to analyse the impacts of institutional arrangements on the supply function of petroleum reserves, to assess how the resulting property rights framework influences the allocation of certain resources, and to obtain an insight into the implications for host country goals of such allocative impacts.

To do so it was necessary to investigate the institutional framework and the variables influencing the supply function. There were three main aspects requiring study: (1) the technical aspects, that is, the geological setting, which provides the basis for a potential supply function; (2) the industry setting within which the contractor or firm would undertake to supply these reserves, and (3) the legal-political and economic setting or environment within which the supplier of these reserves must operate. The allocative impacts on the supply of petroleum reserves of the interaction of these three elements are summarized in this chapter in the context of host country policy goals. Preliminary to doing so, however, a conceptual summary of the property rights framework is presented.

THE INSTITUTIONAL SETTING AND THE FIRM

It was shown that a demand exists for petroleum reserves in the host countries of South-East Asia. It was also established that there were geological indications of potential accumulation of these reserves within the onshore and offshore political boundaries of most of the countries in South-East Asia. Although none of the accumulations was expected to be found in fields akin in size to those of the Middle East, such accumulations were not in any sense negligible and, in fact, were of some significance in the context of the region and of the individual countries.

It was suggested from the description of contractual rights in Chapters III and IV that the production function of the petroleum firm in South-East Asia may not be conceptualized in the conventional sense. The peculiar situation is summarized in the succeeding paragraphs.

1. The basic input in the exploration company's production function is the petroleum-bearing area, and the final output is the supply of petroleum reserves out of the discovered resources. A petroleum company operating in South-East Asia is therefore faced with the following situation. The owner of the basic input in exploration is the state. In any particular country the firm is dealing with a monopoly in terms of the supplier of its basic input. On the output side, there are few or several suppliers of petroleum reserves, but not many in the connotation associated with perfect competition. Thus the situation is very much that of an oligopoly, with the number of producers ranging from a few to many, but not so many that each company perceives that its independent action has an imperceptible effect on the overall outcome (e.g., in dealing with the supplier of its input, the host government, or in supplying reserves to the host government).

2. The situation may be viewed in another light. A review of the contractual terms of the production-sharing contract and its variants (e.g., the service contract) will show that the petroleum firm may not be considered an independent producer in the usual sense. Rather, it is a contractor and sells its services to the host government, doing so to supply that government with petroleum reserves. The firm invests its capital, know-how, and technology in the venture with the host country, in return for which it receives a share of the petroleum output. It also takes the primary 'risk' in exploration. Conceptually, however, the total final output—the petroleum reserves—still belongs to the state. Because the firm has control over the technical aspects of the operation, it can vary the amount of output (or payment) to some extent. But it does not have com-

plete control over the amount of the ultimate supply of the reserves in any particular contract area; its services may be terminated by itself or by the host country, and production from the same acreage would still be possible if the host government engages another contractor to operate the contract area. (The firm, of course, has other options; it may opt to operate under the conditions stipulated by the host government, or it could opt to invest in alternative opportunities.)

3. The pricing of the output once production takes place is not determined in the market-place in the usual sense of the word nor by the firm. For most purposes, the international price of crude oil in South-East Asia is determined by OPEC; that for domestic consumption is often set by the state. (When the market is slack, the Indonesian state marketing firm has acted on its own in pricing international sales, ignoring cartel agreements on prices; in general, however, prices are determined by the cartel. The latter is true, even though a wider range of prices began to appear in the international market in 1979, as a tight market situation developed with the Iranian crisis.)

4. The supplier of petroleum reserves operates within the bounds of conditions 'negotiated' with the owner of the primary input, the petroleum-bearing lands. Given a bargaining situation between a monopoly seller of primary inputs and an oligopolistic buyer of those inputs, a determinate solution is not possible. Rather, the bargaining process and its results are highly dependent on the strengths of the parties involved.

This study thus reveals that the contractual and political framework within which the private firm operates is a very complex one. It is possible to draw a general characterization of the setting, but the specific conditions under which a contract area is operated may differ from firm to firm, from country to country, or from year to year. However, the profit-maximizing motive of the firm—however profit is defined—remains a paramount factor in assessing response to a policy variable. In other words, given the geological potential of a project, the supply of petroleum reserves from that area is in the end heavily dependent on the nature of the property rights arrangements and the concomitant economic returns that a firm can expect to get from a specific operation. The implications of the setting described above are examined in the following sections.

STATE OWNERSHIP AND ATTENUATION OF RIGHTS

What are the effects of state ownership of the most important raw material input in the production function? And what are the effects of the strong regulatory aspects of this ownership?

State ownership implies that the state can exclude anyone from the use of the right, so long as the state follows accepted political procedures for determining who may or who may not use state-owned property. Since the state in South-East Asia is the owner of the petroleum resource, a firm can acquire rights to produce the resource by contractual arrangement with the state, and as pointed out earlier, its property rights are limited to those of a 'hired hand'.

In developing countries, the term 'risk' is increasingly taken to include not only geological and technical risks but also the 'political risk' element. For the purpose of studying South-East Asia, the term 'political risk' might include the sudden changes in rules that attenuate the property rights of the firm as well as those that alter its profit calculi. In computing its expected rate of return, the firm makes assumptions with regard to its revenue and its costs. In the case of the first, it may make any of three assumptions given expected demand: (1) that price in real terms will not change, (2) that prices will drop in real terms as new energy sources are developed, or (3) that prices will increase in real terms, on the basis of recent experience. In computing its costs, it may make an assumption that (1) 'government take' will not change during the recovery period, or (2) that this will increase after any number of years. Whether the first assumption is realistic or not is difficult to say. The recent record for South-East Asia indicates a high degree of instability and unpredictability in the shifts in government policies. The cases of Indonesia, Malaysia, and the Philippines are some examples. The presence of political uncertainty and risks tends to result in underinvestment by firms in exploration and development if firms discount future revenues too heavily. This study suggests that this may be occurring in South-East Asia.

TYPE OF FIRMS AND SOCIAL OPTIMIZATION

On the other hand, how about the industry framework, in terms of type of firms engaged in exploration and development? The optimization behaviour that may be expected from the existing type of firm gives rise to other implications. In planning its resource development strategies, a host government may not ignore the industry's structure.

In the developing countries of South-East Asia, because of the substantial 'risk' capital required, the firm engaged in the first two stages of the industry would, in general, be of foreign origin. The firm may either be a multinational giant or a small international firm. In the early stages, or before the producing capacity of a country or region has been established, the giant would more likely be the firm engaged in exploration and development for reasons discussed in Chapter III. The dominance of international firms implies several conditions influencing the response of firms to stimuli.

As a firm operating outside its home country, the firm is subject to factors both internal and external to those in the host country. For almost any foreign firm, home country policies affecting its profit base must be considered in addition to those in the host country. As a foreign firm also, questions about equity participation in the venture, repatriation of profits, and taxation once production takes place become relevant.

The structure of the operating firm in itself implies several levels of decision making and optimization. The major multinational company calculates its profits not merely in terms of its operations in a particular host country but on a global basis. The decision to invest in that operation is not independent of results from the firm's other operations or from developments elsewhere which may affect the firm's overall returns. That is, while the local manager may compute its costs and profits in terms of the specific contract area, (1) the very nature of petroleum exploration—which draws from gains in projects in other areas—requires an overall assessment by the top management of the firm and a ranking of the profitability of prospects; and (2) by the very nature of its being a multinational company, it will have alternatives to choose from and therefore be able to weigh its opportunity costs. Thus, the decision not to explore in one area may not necessarily be an indication of the economic viability of that project but rather the rank of that prospect in terms of the firm's prospects elsewhere and the most profitable allocation of its investment funds. For example, Atlantic Richfield reduced work in Indonesia to meet commitments in Alaska, as production from its Alaskan fields—an investment with an expected greater stability—became a near-term prospective reality. An offsetting factor may be possible if the firm invests abroad to spread its investments so that it may reduce its risks. The Tobin-Markowitz theory suggests that a low (or negative) correlation between foreign and domestic risks can make investment in exploration and development overseas attractive under those circumstances.[1]

A practical example suggests that the notion of profit maximization may, in fact, be practised by an MNC on a global rather than on an area scale. An exploration manager for a U.S. multinational company in South-East Asia indicated to the author that, having made an initial investment under given marketing prospects, his company was willing to operate within the minimum profit allowed by the contractual and fiscal framework—but only up to a certain point. The notion of marginalism is not necessarily rejected in this example, of course, if one can quantify the probability of failure or success and the costs of waiting over and above the cost of capital that is actually involved (the opportunities forgone)—especially if such calculi cover the MNC's total operation.

On the part of the international firm, therefore, given the same working conditions, it can opt to drill and develop the more attractive potential, that is, the field with the larger reservoir under given cost conditions and availability of funds. Alternatively, given the same size and type of prospects, it can opt to invest its funds in that country with the more attractive working conditions.

The presence of such options for the international firm places a 'floor' on the control of the host government over the level of exploration investments it can attract, given both technical and economic conditions related to its resources. In addition, a problem arises where the government's technical knowledge is insufficient.

When dealing with large MNCs, whose alternative choices span the range of the globe, a host government in a developing country can be faced with a dilemma. How does it judge that retrenchment by an oil company in the face of what it claims are 'unfavourable terms' is a real retreat—the effect of legislative or contractual disincentives—or if it is merely a threat by the firm perceiving that it has sufficient leverage and a relatively strong bargaining power? (The data suggest, through similar reactions by several MNCs, that the former may have been the case in South-East Asia.) This is only one of many difficult questions in government-company relations. They certainly have serious implications for overall resource allocation in the region.

THE CRUCIAL SUPPLY VARIABLES AND ALLOCATIVE IMPLICATIONS

What then are the crucial supply variables? Actually they still boil down to *revenues* and *costs* with risks entering the cost category. Thus, a simple model may be represented by showing the supply of reserves as a function of the supply of wildcats drilled, given the geological prospects. And the supply of wildcats may be viewed as a function of the factors entering the investment decision, more specifically a function of the contract terms and other non-area variables. If we take the non-area variables as given, then the contract terms will determine the supply of wildcats.

The firm's *revenues* over its investment planning period may be viewed in terms of the size of the discoveries and price levels. The study showed that, as might be expected, drilling activity responded to the economic incentives represented by the size of discoveries and the success ratio. That many of the smaller companies have been willing to go offshore, despite the more costly risks, may indicate that the geological prospects are better, their knowledge of the area is better, or simply that the successes of others have provided them with the economic incentive to try their luck, whether or not they know what they are doing. This study has, after all, shown that discoveries are random and, to some extent, must be attributed to luck. Because of this, it was possible for both major and relatively small companies to chance upon large discoveries, such as the finds in Indonesia by ARCO (Ardjuna), IIAPCO (Cinta), and Union (Attaka).

Costs include both technical and fiscal costs. Neither the firm nor the host government (or supplier of the primary input) has much control over the technical costs. These may be taken as given for purposes of this study. They are determined by the geological environment of the wells and technology. That is, costs are related to the dry-hole ratio, well depth, water depth, location in relation to services and supplies that affect logistics (e.g., frontier areas). Of course, a firm can do more or less work, but it is safe to assume that some minimum is both technically and financially desirable. The only aspect of costs that are within the control of either party, assuming equilateral bargaining is possible, are the fiscal costs in the form of taxes, royalties, and production shares. To the extent that the host country holds the upper hand, it has relatively greater control over the level of fiscal costs and therefore can affect the cost-reward structure for the firm.

Van Meurs used the notion of *conditions pressure* (*CP*) in analysing a firm's response to legislation.[2] By raising one or more types of payments to the government the conditions pressure becomes higher, and vice versa. Incorporating the presence of geological risk, the expected monetary value of an investment is used as a yardstick for measuring this pressure, expressed as follows:

$$CP = \frac{EMV - EMV'}{EMV}$$

where *EMV* is the expected monetary value of a project if no payments to the government are made; and *EMV'* is the value when payments are required. When *CP* = 1 (i.e., when *EMV'* = 0), the company is barely willing to take the risk to explore.[3]

An equally important issue is that of *risk*. The manner in which the burden of exploration risk is allocated affects the choice of locations to drill as well as the type of contractors that may seek to drill. The greater the burden that must be borne, assuming a risk-averse exploration firm, the greater will be the tendency to choose the sites that involve either relatively lower risks or greater chances of success. This could mean that there is probably relative under-investment in the untested areas and relatively lower exploration of

frontier areas or even some offshore areas. The surge in new contracts in 1979 does not change this conclusion—the contracts are
for areas where previous work (with favourable results) had been
done.[3a]

For example, the signature bonus as a method of extracting
revenues from the contractors may appear to be an optimal way,
from the government's viewpoint, of extracting rent from the producer. It is sometimes argued by host country officials that companies pay 'what they can bear' or what they perceive is the value
of the right to explore and possibly to develop an area. On the
other hand, if the argument is correct that bonus bids are sharply
discounted for risk, then it is possible that the optimal number of
contractors is not attracted under the present system.[4] Also, it may
be argued that bonus bids tie up capital that would otherwise be
free for immediate exploration and development and thus retard
access by the host government to the benefits from such activities.

At the same time, as noted earlier, the cost recovery framework
of the production-sharing contract would appear to result in
under-investment in the relatively higher-risk areas. The approach
embodied in this framework is contrary to the approach that has
been recommended in other areas, such as the United States. One
study on petroleum leasing policies in the outer continental shelf
(OCS) of the U.S. specifically recommended that optimal resource
development may require that the Federal government bear some
of the risks of exploration and provide greater pre-drilling information to prospective lessors.[5] (This assumes, of course, that OCS
development is desirable.) Game theoretic models have indicated
that the government's share of economic rent increases in some
direct proportion to the amount of information an interested company has on a prospective contract area.[6] However, more information on many untested areas in South-East Asia is not forthcoming
without drilling, and drilling will not be undertaken in high-risk
areas if losses are not recoverable in some form from a company's
other operations. It is a vicious cycle, but it probably is one of
those difficult pre-conditions to exploration investment in frontier-
type areas that requires some serious considerations and trade-offs
in policy strategies.

The property rights approach to explaining the operating conditions of the firm states that the entrepreneur possesses basic ownership rights, one of which is the right to receive the residual after all inputs have been paid contractual amounts. This condition is not violated by the transfer to the state of ownership over petroleum resources. When a state contracts exploration and production to a private firm, however, two levels of ownership are involved—both of which seek to maximize returns but whose objective functions are not necessarily coincident.

The history of exploration in South-East Asia indicates that, given the resource potential and the appropriate technology, economic incentives have determined the level of exploration activity in the area. It is probably useful to go back over a decade to the Philippine example. The 1977 discovery off Palawan Island is in an area that had as early as 1965 been identified as geologically prospective for accumulation by a major oil company. That company withdrew its interests in the area, however, because it failed to get the required assurance of what it perceived to be reasonable long-term investment conditions.

This brings us to the question of what a reasonable return is. Governments have tended to take it upon themselves to decide what is reasonable in light of their knowledge of the company's operating costs. It may know the actual costs incurred in the operations within its jurisdictions, but it has no cost information on the rest of the company's operations. To a large extent then it must rely on the firm's own judgements of what is reasonable. The subject of what is reasonable may be debated *ad nauseum* fruitlessly. In the end, we fall back on the conceptual and measurement problems of what is optimal, where the firm's managers must not only fulfill their own objectives but must yield profits that are 'acceptable' to its owners, the stockholders. A government may unknowingly find itself in the role of pushing a firm beyond what the latter considers optimal.

In fact, shifts in government policies may not necessarily be the result of a bargaining process between the state's agent and the firm but may arise from a state's notion of its bargaining advantage, whether correctly perceived or not. In developing countries, there may be few, if any, persons who know enough about the industry to evaluate the data presented to them by the prospective operators, or to know how correct the industry's claims are. An Indonesian official, for example, recently argued that one of the bases for changing the contractual conditions in the 1970s was that the Indonesian negotiators progressively know more about the industry than they did when they entered into contracts with foreign oil companies in the 1960s. A strong nationalistic sentiment or even the attitude of a state towards the potential earnings to be gained from the development of its resources affects the climate for investment; the problem arises in the perceptions each party holds concerning its strengths.

Game-type strategies are not in themselves undesirable; if the results are positive-sum, both parties gain from playing the game.[7] Achievement of the optimum situation is also possible if the two actors in this bargaining process, although rivals, recognize their interdependence and move towards some area of coincidence. This may be expressed conceptually as somewhere between the two vectors representing each actor's objectives (see Figure 7.1), or in terms of the Edgeworth *contract curve*.[8]

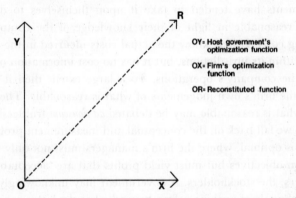

OY= Host government's optimization function

OX= Firm's optimization function

OR= Reconstituted function

Figure 7.1. Reconstitution of Optimization Functions

(With apologies to Pierre Teilhard de Chardin's diagram of the conflict between the two kinds of faith in man's heart, *The Future of Man* (New York: Harper and Row, 1964, p. 282.)

*OR is not necessarily a half-measure but a reconstitution of goals to complement each other's goals.

There is nothing wrong with a government's taking it upon itself to decide how much a firm must make within its territory and how much of the economic rent resulting from the firm's operations must be shared with the host country. As stated earlier, the host government also has its own goals. What we are trying to say perhaps is that a host country may need to improve its information level if it is to be successful in its strategies of obtaining the optimum revenues from the development of its resource-bearing lands, and not employ strategies that are counter-productive.

Related to this is the timing of such strategies. This has serious implications for an industry which requires long lead times between the time it makes its investments and the final year for recovering returns. Having calculated its costs and profits to cover a number of years, a change in one of the variables affecting its profit calculations in a project would yield different results. If this change is negative, and if the firm perceives that changing policies rather than stable policies over the life of its future investments is the norm, it could rank future projects in the area lower than another where it can reasonably expect greater stability. Sudden shifts have occurred in South-East Asia in several instances in the last few years. Changes in the institutional framework, whether internal or external to the region, were thus shown in this study to have an impact on the supply of petroleum reserves in the region. By altering the conditions under which revenue/cost relationships were estimated over the investment period, these changes influenced the level of exploratory drilling. Perhaps more important and more serious than the short-term reaction is the long-term response of other foreign investors. A record of sudden shifts in government policies could result in higher 'risk factors' in ranking rates of returns.[9]

OTHER ALLOCATIVE AND POLICY IMPLICATIONS

The impacts of these shifts affect the long-term reserve picture of producing countries. They also extend to the rate of lifting in existing projects. For example, changes in 1976 in the operating conditions and cost/reward structure in Indonesia not only resulted in a perceptible drop in new entrants; they also raised questions regarding the rate at which existing fields would be produced as well as the rate at which existing discoveries would be developed.

It might be useful at this point to pause and review an illustrative situation—the petroleum reserve situation in Indonesia over the period 1972 to 1978. Table 7.1 shows that the net additions to reserves during the period have been at a compounded rate of only 0.4 per cent, while production grew at a compounded growth rate of 8.2 per cent. A significant addition to the country's reserves occurred only in 1974, and in the three subsequent years reserves were drawn down. It is also useful to note that while total reserves

TABLE 7.1

INDONESIA: OIL RESERVE POSITION AND PRODUCTION DATA, 1972–1978

	Oil Reserves (10^9 bbl.) (as of 31 Dec.)	Production (10^3 b/d) (as of 31 Dec.)	No. of Wells Producing (as of 1 July)	Output per Well (b/d)
1972	10.0	1,027	2,344	438.1
1973	10.5	1,300	2,567	506.4
1974	15.0	1,457	2,710	537.6
1975	14.0	1,300	3,018	430.8
1976	10.5	1,500	3,162	474.4
1977	10.0	1,690	3,421	494.0
1978	10.2	1,650	3,644	452.8
	Net addition to reserves (10^9 bbl.)	% Δ	% Δ	% Δ
1973	+0.5	+26.6	+9.5	+15.6
1974	+4.5	+12.1	+5.6	+6.2
1975	−1.0	−10.8	+11.4	−19.9
1976	−3.5	+15.4	+4.8	+10.2
1977	−0.5	+12.7	+8.2	+4.1
1978	+0.2	−2.4	+6.5	−8.3
Compounded growth rate	+0.4	+8.2	+7.7	+0.6

Source: Basic data from Oil and Gas Journal, year-end issues.

in 1977 and 1978 were at or just slightly above the level of 1972, the number of wells producing had grown by over 150 per cent, with a concomitant increase in the total daily output. In 1978, therefore, the 10.2 billion barrels of reserves had a remaining life of only 17 years at the 1977 rate of output, whereas in 1972 a 10 billion-barrel reserve had 27 years remaining at the 1972 rate of output.

These patterns suggest several things. The record of the years 1976 to 1978 may indicate that companies have been going for smaller but relatively less risky prospects, and have been staying away from the untested areas. (As stated earlier, of course, companies think that the only prospects left in Indonesia are relatively small fields.[10] How this thinking is influenced by the contractual framework is not clear.) The severe drawing down of reserves in 1976 also reflects the effects of the contract renegotiations that started with Caltex in 1975. The drop in output per well below the 1973 and 1974 levels, despite the hefty price increases that followed the 1973 OPEC embargo, appear to support theory.[11]

In addition, the patterns suggest that companies may be developing their contract areas relatively more intensively. It is possible that the future is being discounted rather more heavily (see the discussion in Chapter III again) and that firms are probably trying to recover their investments in a relatively shorter time frame. (This has other implications with respect to the cost of production, but without additional information we will not venture any guesses.) It is generally acknowledged that state policies affect the private discount rate such that they alter the private firm's perception of the efficient rate of lifting. Such rate of extraction may not necessarily be the most efficient rate of recovery in engineering terms—which normally would be the societal criterion for efficiency from the standpoint of conservation. It has been suggested on occasion that foreign firms may tend to produce at relatively faster flow rates in developing countries or in areas outside their home countries. It is not clear, of course, if these flow rates exceed the 'maximum efficient rate' or MER. It is quite reasonable to expect that no producer would want to damage the reservoir from which he is currently extracting oil to the point that his total

returns would be seriously impaired.[12] The possibility, nevertheless, exists that a firm defines the 'optimum' yield or MER differ-
ently in different investment climates.

On the other hand, the data may include some dampening
effects if a firm were to produce at points other than the 'optimal'
point in the reverse direction; that is, firms may produce below
what might be optimal to evade payment of a bonus. Any non-
production cost that can be made to change with changes in production could bias production. Thus a possible reaction to the production bonus, or government 'take' that is tied to certain output
levels, could limit output below the optimal point. This issue is
recognized by some host government officials who concede that
certain firms which are producing at levels close to that subject to a
bonus may tend to 'underproduce'. A firm may need a substantial
price increase before it considers the increased output (and resulting revenues) sufficient to cover payment of the bonus without
impairing its desired profit-margin. In this case any increase in
world prices may not necessarily result in increased output, with
such output rising only at substantial price increases. Thus the
supply curve may look somewhat like the kinked curve in Figure
7.2.

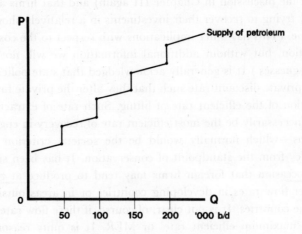

Figure 7.2 The Kinked Supply Curve: A Possible Response
of the Firm to the Production Bonus

A practical example might help show that it is very possible for a host government to lose revenues unwittingly as a result of this bonus provision, if a company decides to evade paying such bonus by producing below, say, the bonus level of 50,000 barrels per day. At the January 1979 price of US$13.90 per barrel of Indonesian oil, a company in Indonesia producing at 46,000 barrels per day will receive the same net output as at 50,000 barrels per day over a 120-day period after paying a US$1 million bonus and giving the government its 85 per cent share.[13] Thus, it could produce at 48,000 barrels per day and be receiving a higher net output share. On the other hand, because the company is not producing at 50,000 barrels per day, the government is losing revenues. If the company produces at 48,000 b/d, the government loses over US$2.8 million per 120-day period in terms of its 85 per cent share. This should not be of major concern if the host government does not want the petroleum or the revenues from the petroleum at this time but would like to conserve its resources for future generations (e.g., Malaysia's claims). Both Indonesia and the Philippines do appear to want to extract the resource or the revenues in the immediate or near-term, and may, in that context, be finding their production bonus provisions somewhat counter-productive.

While these last few points refer to the later stage of the upstream phase of the industry, these are considerations that undoubtedly do not escape a firm in its decisions to engage in exploration for petroleum reserves. Probably more serious over the long-term is the effect of these contractual arrangements on the rate at which South-East Asia's petroleum resources will be developed. As noted earlier, Grossling's approach to the study of the petroleum potential in developing countries indicates that the prospective areas in South-East Asia, among others, have hardly been tapped. Let us take Indonesia's example again. Table 2.2 shows Indonesia's potential recoverable resources to be in the range of 10 to 100 billion barrels. Cumulative production[14] and proved reserves as of the end of 1978 amounted to about 18 billion barrels, indicating a 'high' recoverable potential of about 80 billion barrels remaining to be discovered. If a conservative undiscovered potential of 50 billion barrels is assumed, this would mean around

80 additional years of production at 1978 levels remaining to be discovered—over and above 1978 proved reserves. It is, of course, up to the present generation and its government to decide how it wants to continue developing its petroleum resources and at what rate it wants additions to its reserves. A host government's strategies—and their effectiveness—will depend on the agent it chooses to develop those resources.

CONCLUDING COMMENTS

The analyses have tended to focus on the production-sharing contract and its variants. This is because this form is dominant in most of the countries in South-East Asia where exploration is being actively pursued. It is also a form that appears to be increasingly accepted as an optimal way of incorporating a host country's socio-economic goals related to its petroleum resource policies. Even where the concession is in force, however, government control is effectively implemented in a more forceful way than in earlier periods, and the explicit contractual terms are mere deviations from the general pattern of host country control.

It has also been assumed all along in these discussions that the governments of South-East Asia want to develop their petroleum resources. The definition of the optimum time frame appears to rest largely on the social rate of discount that a government places on the extraction of those resources. This rate may vary from country to country. Whatever the case may be, we may sum this all up as follows:

The property rights arrangements covering petroleum resource development in South-East Asia imply certain economic responses on the part of the firms contracting to develop such resources. The data indicate that the supply of such petroleum reserves is in fact sensitive to such arrangements. The structure of the industry also suggests that decisions to supply such reserves are generally made within a context that extends beyond the boundaries of the region. Analyses of the specific contractual terms governing costs recovery and production bonuses suggest, however, that, where the goal of the host country is accelerated development of its petroleum resources, there may in fact be relative underinvestment in the exploration and development of such resources, especially in high-risk areas. These contract provisions have serious implications for the supply of petroleum reserves in South-East Asia. They suggest that a better understanding of a firm's behaviour may be required to allow a host government to develop strategies that will optimize its supply of reserves as well as the benefits it wants to capture from the development of its resource endowments.

1. See the example given in Benjamin Cohen, *Multinational Firms and Asian Exports* (New Haven and London: Yale University Press, 1975), p. 25. Cohen notes the following:

Let x be the variable—such as sales or profits—on which management focuses. Let p be the proportion of x in country 1 and $1 - p$ the proportion in country 2. Let r be the correlation between x in country 1 and x in country 2. Let v_{1+2}^2 be the total variance of x, v_1^2 the variance of x in country 1, and v_2^2 the variance of x in country 2. Then:

$$v_{1+2}^2 = p^2 v_1^2 + 2p(1 - p)rv_1 v_2 + (1 - p)^2 v_2^2$$

For example, suppose the variance of sales is nine in one country and sixteen in another country and the correlation between sales in the two countries is .25; then having half of a firm's sales in each country gives a variance for the firm's worldwide sales of 7.75. So firms in both countries reduce their variance through foreign investment.

2. Van Meurs, *Petroleum Economics and Offshore Mining Legislation*, pp. 104–5. The term *conditions pressure* is a notion used by Van Meurs to refer to the burden felt by companies in reacting to payments required by a host government, that is, to the fiscal conditions.

3. Van Meurs' basic model is expressed in terms of the net-present value. The conditions pressure in terms of the net-present value is defined as follows: $CP = \dfrac{NPV - NPV'}{NPV}$, where NPV is the net-present value without government 'take' and NPV' is the net-present value where government 'take' is paid. When no payments are made to the government, NPV' is equal to NPV and conditions pressure is equal to zero (that is, $NPV' = NPV$, so that $NPV - NPV' = 0 = CP$). When the net-present value of a venture is zero after payments to the government are made, the conditions pressure is equal to 1, and the project is marginal to the producer. The same analysis applies to the expected monetary value with government 'take', EMV'.

3a. See *PE*, January 1980, pp. 31–2, 'Rush for exploration contracts'.

4. See Richard B. Norgaard, 'Uncertainty, Competition, and Leasing Policy', unpublished report submitted to the State of Alaska (Berkeley: University of California, 1977).

5. See Leland *et al.*, 'An Economic Analysis of Alternative Outer Continental Shelf Petroleum Leasing Policies'.

6. See Norgaard, op. cit., p. 10.

7. The reader who is not familiar with the basics of game theory may refer to William J. Baumol, *Economic Theory and Operations Analysis* (Englewood-Cliffs: Prentice-Hall, 1961), Chapter 18.

8. The Edgeworth contract curve joins the points of tangencies of the indifference (or trade-off) curves of two negotiating parties, in this case the firm and the government. The equation for an indifference curve may be expressed as follows:

$$\frac{\partial U}{\partial x} \, dx + \frac{\partial U}{\partial y} \, dy = 0,$$

where U = the utility (satisfaction) of each actor, x = gains, and y = losses or costs.

At the points of tangencies the slopes of each actor's indifference curves $(-dy/dx)$ are equal. That is, their trade-off rates are equal. At this point also, the two parties are at mutually beneficial positions.

9. In the Rummel and Heenan article, 'How Multinationals Analyze Political Risk', Indonesia and Malaysia were mentioned by a multinational oil company executive as 'borderline cases' in its assessment of countries with very stable and very unstable investment environments.

10. This does not mean that numerous large and important discoveries will not be made in Indonesia over the coming years; the 'giant' fields, however, probably may not be forthcoming.

11. Economic theory suggests that output will be reduced if marginal cost is increased with price held constant. The effect of the US$1.00 per barrel tax on Caltex was to increase the company's marginal cost. Increase in the government's share of production has a similar effect, in so far as such 'take' increases a company's costs.

12. The last term in Equation 3.4 in Chapter III is unique to oil production and reflects the fact that high rates of production today can actually reduce the total amount of oil that can ultimately be recovered. See Lovejoy and Homan, *Economic Aspects of Oil Conservation Regulation*, p. 92.

13. This assumes 'cost oil' has already been excluded.

14. See *OGJ*, year-end issue.

Appendixes

APPENDIX A

Glossary of Terms in the Petroleum Industry

API gravity: A parameter, expressed in degrees, and mathematically related to specific gravity. When expressed as API gravity (API stands for American Petroleum Institute), water has a specific gravity of 10.00 degrees. The API scale is: degrees hydrometer scale (at 60°F) = (141.5/sp. gr.)—131.5. The lighter the crude, the higher the API gravity. The gasoline and kerosene content of crude oil tends to be directly related to the API gravity.

Commercial accumulation: An occurrence of oil and gas that meets the minimum requirement for size and accessibility to be of commercial interest to a company. The term 'commercial' is frequently synonymous with 'economic'.

Deliverability: The amount of natural gas that a well, field pipeline, or system can supply in a given period of time. Only valid for that period.

Discovered resources: That portion of the oil and gas in the earth whose presence has been physically confirmed through actual exploration drilling.

Indicated reserves: Known oil and gas that is currently producible but cannot be estimated accurately enough to qualify as proved.

Inferred reserves: Reserves that are producible but the assumption of their presence is based upon limited physical evidence and considerable geologic extrapolation. This places them on the borderline of being undiscovered. The accuracy of the estimate is very poor.

Inplace: All of the oil and gas in the reservoir, combining both the recoverable and non-recoverable portions.

Maximum efficient rate (MER): When used in a practical or operational sense, it is the optimum rate, as of a specific time, at which oil and gas should be drawn from a developed field in order to balance cost, percentage recovery, and speed of withdrawal. To exceed this rate for the reservoir or to produce individual wells too rapidly can lead to loss of oil and gas recovery from the reservoir.

Occurrence: A physical accumulation of oil or gas or related hydrocarbons in the earth regardless of size and physical or economic characteristics.

Oil basin: A large basin-like geologic structure in which oil and gas fields will be found.

Oil field: A geologic unit in which one or more individual, structurally and geologically related, reservoirs are found.

Oil region: A large oil-bearing area, often encompassing several states, in which oil basins and fields are found in close proximity.

Production or decline curve (S curve): The annual production of an oil or gas reservoir through time is a dome-shaped profile with its peak usually to the left of centre. The progress of the production from its peak toward depletion is called the 'decline curve'. If this is plotted as cumulative production it follows a gradual S-shape as it approaches the total, or ultimate, production of the reservoir.

Productive capacity: The amount of oil that can be withdrawn each day from existing wells with available production facilities. Only valid at one point in time.

Proved reserves: An estimate of oil and gas reserves contained primarily in the drilled portion of fields. The data to be employed and the method of estimation are specified so that the average error will normally be less than 20 per cent. May also be called measured reserves.

Recoverable: That portion of oil and gas resources that can be brought to the surface, as distinct from the oil and gas found in place in the reservoir.

Reserves: Oil and gas that has been discovered and is producible at the prices and technology that existed when the estimate was made.

Reservoir: A continuous, interconnected volume of rock containing oil and gas as a hydraulic unit.

Resource base: The total amount of oil and gas that physically exists in a specified volume of the earth's crust.

Resources: The total amount of oil and gas, including reserves, that is expected to be produced in the future.

Sub-economic resources: Oil and gas in the ground that are not producible under present prices and technology but may become producible at some future date under higher prices or improved technology.

Undiscovered resources: Resources which are estimated totally by geologic speculation with no physical evidence through drilling available.

Source: John J. Schanz, Jr., 'Oil and Gas Resources—Welcome to Uncertainty', in *Resources*, No. 58 (Washington, D.C.: Resources for the Future, March 1978); Van Nostrand's *Scientific Encyclopedia*, 5th ed. (New York: Van Nostrand Reinhold, 1976).

APPENDIX B

Parts of an Actual Production-sharing Contract

SECTION II
TERM

1.1 The term of this Contract shall be thirty (30) years as from the Effective Date.

1.2 If at the end of the initial six (6) years as from the Effective Date no Petroleum is discovered in the Contract Area, CONTRACTOR shall have the option either to terminate this Contract or to request PERTAMINA by means of a thirty (30) days' written notice prior to the end of the initial six (6) years' period to extend the aforesaid six (6) years' period for two additional periods of two (2) years each which extensions shall be promptly granted, without prejudice to the provisions of Section III regarding exclusion of areas and Section XIV relating to termination.

1.3 If after the initial six (6) years' period or any extensions thereto, as from the Effective Date, no Petroleum is discovered in the Contract Area, this Contract shall automatically terminate in its entirety.

1.4 If petroleum is discovered in any portion of the Contract Area within the initial six (6) years' period, or any extensions thereto, which in the judgment of PERTAMINA and CONTRACTOR can be produced commercially, based on consideration of all pertinent operating and financial data, then as to that particular portion of the Contract Area development will commence. In other portions of the Contract Area explorations may continue concurrently without prejudice to the provisions of Section III regarding the exclusion of areas.

SECTION III
EXCLUSION OF AREAS

1.1 On or before the end of the initial five (5) years' period as from the Effective Date, CONTRACTOR shall surrender twenty-five percent (25%) of the original Contract Area.

1.2 On or before the end of the eighth Contract Year CON-
TRACTOR shall surrender an additional area equal to
twenty-five percent (25%) of the original total Contract Area.

1.3 On or before the end of the tenth Contract Year, CONTRAC-
TOR shall surrender such additional area that the area
retained thereafter shall not be in excess of Three Thousand
(3,000) square kilometers.

1.4 CONTRACTOR'S obligations to surrender parts of the original Contract Area under the preceding provisions shall not
apply to any part of the Contract Area corresponding to the
surface area of any field in which Petroleum has been discovered.

1.5 With regard to the remaining Contract Area left after the
above mandatory surrenders, PERTAMINA and CON-
TRACTOR shall maintain a reasonable exploration effort. If
in respect of any part of such remaining Contract Area CON-
TRACTOR does not during two (2) consecutive years submit
an exploration program, PERTAMINA shall reach agreement
with CONTRACTOR as to whether any such part of the Area
should be surrendered or whether exploration thereon should
be resumed at a later date.

1.6 Upon thirty (30) days' written notice to PERTAMINA prior
to the end of the second Contract Year and prior to the end of
any succeeding Contract Year, CONTRACTOR shall have
the right to surrender any portion of the Contract Area, and
such portion shall then be credited against that portion of the
Contract Area which CONTRACTOR is next required to surrender under the provisions of subsections 1.1, 1.2 and 1.3
hereof.

1.7 CONTRACTOR shall advise PERTAMINA in advance of the
date of surrender of the portion to be surrendered for the purpose of such surrenders. CONTRACTOR and PERTAMINA
shall consult with each other regarding the shape and size of
each individual portion of the areas being surrendered; provided, however, that so far as reasonably possible, such portions shall each be of sufficient size and convenient shape to
enable Petroleum Operations to be conducted thereon.

SECTION IV
WORK PROGRAM AND EXPENDITURES

1.1 CONTRACTOR shall commence Petroleum Operations hereunder no later than six (6) months after the Effective Date.

1.2 The amount to be spent by CONTRACTOR in conducting Petroleum Operations pursuant to the terms of this Contract during the first eight (8) years following the Effective Date shall in the aggregate be not less than hereafter specified for each of these eight (8) years as follows:

First Contract Year	U.S.$1,400,000
Second Contract Year	U.S.$1,300,000
Third Contract Year	U.S.$1,300,000
Fourth Contract Year	U.S.$1,200,000
Fifth Contract Year	U.S.$1,200,000
Sixth Contract Year	U.S.$1,200,000
Seventh Contract Year	U.S.$1,200,000
Eighth Contract Year	U.S.$1,200,000

provided that at any time following the expiration of three (3) years after the Effective Date, and provided CONTRACTOR has complied with its obligations under this subsection 1.2 of Section IV applicable through the then current year, CONTRACTOR shall have the right to relinquish all of the Contract Area, whereupon these expenditure obligations relating to years subsequent to such relinquishment shall terminate. If during any Contract Year CONTRACTOR shall expend more than the amount of money required to be so expended by CONTRACTOR, the excess may be subtracted from the amount of money required to be expended by CONTRACTOR during the succeeding Contract Years.

1.3 At least three (3) months prior to the beginning of each Contract Year or at such other times as otherwise mutually agreed by the Parties CONTRACTOR shall prepare and submit for approval to PERTAMINA a Work Program and Budget for the Contract Area setting forth the Petroleum Operations which CONTRACTOR proposes to carry out during the ensuing Contract Year.

1.4 Should PERTAMINA wish to propose a revision as to certain specific features of said Work Program and Budget, it shall within thirty (30) days after receipt thereof so notify CONTRACTOR specifying in reasonable details its reasons therefor. Promptly thereafter, the Parties will meet and endeavor to agree on the revisions proposed by PERTAMINA. In any event, any portion of the Work Program as to which PERTAMINA has not proposed a revision shall in so far as possible be carried out as prescribed herein.

1.5 It is recognized by the Parties that the details of a Work Program may require changes in the light of existing circumstances and nothing herein contained shall limit the right of CONTRACTOR to make such changes, provided they do not change the general objective of the Work Program.

1.6 It is further recognized that in the event of emergency or extraordinary circumstances requiring immediate action, either Party may take all actions it deems proper and advisable to protect their interests and those of their respective employees and any costs so incurred shall be included in the Operating Costs.

1.7 PERTAMINA agrees that the approval of a proposed Work Program will not be unreasonably withheld.

SECTION V
RIGHTS AND OBLIGATIONS OF THE PARTIES

1.1 Subject to the provisions of paragraphs f), g) and h) of subsection 1.2 of this Section V:

1.2 CONTRACTOR shall:

a) advance all necessary funds and purchase or lease all material, equipment and supplies required to be purchased or leased with Foreign Exchange pursuant to the Work Program;

b) furnish all technical aid, including foreign personnel, required for the performance of the Work Program, payment whereof requires Foreign Exchange;

c) furnish such other funds for the performance of the Work Program that requires payment in Foreign Exchange, including payment to foreign third parties who perform services as a contractor;

d) be responsible for the preparation and execution of the Work Program which shall be implemented in a workmanlike manner and by appropriate scientific methods, and CONTRACTOR shall take the necessary precautions for protection of navigation and fishing and shall prevent extensive pollution of the sea or rivers. It is also understood that the execution of the Work Program shall be exercised so as not to conflict with Government obligations imposed on the Government by International Law;

e) retain control to all leased property paid for with Foreign Exchange and brought into Indonesia, and be entitled to freely remove same therefrom;

f) have the right to sell, assign, transfer, convey or otherwise dispose of all its rights and interests under this Contract to any Affiliated Company without the prior written consent of PERTAMINA, provided that PERTAMINA shall be notified in writing of same beforehand;

g) have the right to sell, assign, transfer, convey or otherwise dispose of any part of its rights and interests under this Contract to parties other than Affiliated Companies with the prior written consent of PERTAMINA which consent shall not be unreasonably withheld;

h) have the right to sell, assign, transfer, convey or otherwise dispose of all of its rights and interests under this Contract to parties other than Affiliated Companies with the prior written consent of PERTAMINA and the Government of the Republic of Indonesia, which consent shall not be unreasonably withheld;

i) have the right of ingress to and egress from the Contract Area and to and from facilities wherever located at all times;

j) have the right to use and have access to, and PERTAMINA shall furnish all geological, geophysical, drilling, well, production and other information held by PERTAMINA or by any other governmental agency or enterprise, relating to the Contract Area including well location maps;

k) have the right to use and have access to, and PERTAMINA shall make available, so far as possible, all geological, geophysical, drilling, well, production and other information now or in the future held by it or by any other governmental agency or enterprise relating to the areas adjacent to the Contract Area;

l) submit to PERTAMINA copies of all such original geological, geophysical, drilling, well, production and other data and reports as it may compile during the term hereof;

m) prepare and carry out plans and programs for industrial training and education of Indonesians for all job classifications with respect to operations contemplated hereunder;

n) have the right during the term hereof to freely lift, dispose of and export its share of Crude Oil, and retain abroad the proceeds obtained therefrom;

o) appoint an authorized representative for Indonesia with respect to this Contract, who shall have an office in Jakarta;

p) after commercial production commences, fulfill its obligation towards the supply of the domestic market in Indonesia. CONTRACTOR agrees to sell and deliver a portion of the share of the Crude Oil to which it is entitled pursuant to subsection 1.3 of Section VI calculated for each year as follows:

(i) multiply the total quantity of Crude Oil to be supplied by a fraction the numerator of which is the total quantity of Crude Oil produced from the Contract Area and the denominator is the entire Indonesian production of Crude Oil of all petroleum companies;

(ii) compute twenty-five percent (25%) of the total quantity of Crude Oil produced from the Contract Area;

(iii) multiply the lowest quantity computed either (i) or (ii) by a fraction the numerator of which is the quantity of Crude Oil to which CONTRACTOR is entitled under subsection 1.3 of Section VI and the denominator is the total quantity of Crude Oil allocable under the said subsection.

the quantity of Crude Oil computed under (iii) shall be the maximum quantity to be supplied by CONTRACTOR in any year pursuant to this paragraph p). The price at which such Crude Oil shall be delivered and sold hereunder shall be 20 U.S. cents per barrel f.o.b. point of export. CONTRACTOR shall not be obligated to transport such Crude Oil beyond the point of export but upon request CONTRACTOR shall assist in arranging transportation and such assistance shall be without cost or risk to CONTRACTOR;

q) give preference to such goods and services which are produced in Indonesia or rendered by Indonesian nationals, provided such goods and services are offered at equally advantageous conditions with regard to quality, price, availability at the time and in the quantities required.

1.3 PERTAMINA shall:

a) have and be responsible for the management of the operations contemplated hereunder; however, PERTAMINA shall assist and consult with CONTRACTOR with a view to the fact CONTRACTOR is responsible for the Work Program;

b) except as provided in Section VI, assume and discharge all Indonesian taxes of CONTRACTOR including transfer tax, import and export duties on materials, equipment and supplies brought into Indonesia by CONTRACTOR, its contractors and subcontractors, exactions in respect of property, capital, net worth, operations, remittances or transactions including any tax or levy on or in connection with operations performed hereunder by CONTRACTOR its contractors or its subcontractors. CONTRACTOR shall be obligated to pay any sales taxes under laws and regulations in effect on the date this Contract is signed by the Parties, but at rates not exceeding five percent (5%) of sales price or value, it being understood that materials, equipment and supplies brought into Indonesia and used in the operations will not be subject to sales tax; and PERTAMINA shall be obligated to pay sales taxes only to such extent, if any, as they may be in excess of those described above. PERTAMINA shall not be obliged to pay taxes on tobaccos, liquor and personnel income tax; and income tax and other taxes not listed above of contractors and subcontractors. The obligations of PERTAMINA hereunder shall be deemed to have been complied with by the delivery to CONTRACTOR within one hundred twenty (120) days after the end of each Calendar Year of documentary proof in accordance with the Indonesian fiscal laws that liability for the abovementioned taxes has been satisfied, except that with respect to any of such liabilities which CONTRACTOR may be obliged to pay directly, PERTAMINA shall reimburse it within sixty (60) days after receipt of invoice therefor. PERTAMINA should be consulted prior to payment of such taxes by CONTRAC-

TOR or by any other party on CONTRACTOR'S behalf;

c) otherwise assist and expedite CONTRACTOR'S execution of the Work Program by providing facilities, supplies and
personnel including, but not limited to, supplying or otherwise
making available all necessary visas, work permits, transportation, security protection and rights of way and easements as
may be requested by CONTRACTOR and made available
from the resources under PERTAMINA'S control. In the
event such facilities, supplies, or personnel are not readily
available, then PERTAMINA shall promptly secure the use of
such facilities, supplies and personnel from alternative sources.
Expenses thus incurred by PERTAMINA at CONTRAC-
TOR'S request shall be reimbursed to PERTAMINA by
CONTRACTOR and included in the Operating Costs. Such
reimbursements will be made in United States dollars computed at the rate of exchange extended by the Indonesian Government to Petroleum Companies at the time of conversion.
CONTRACTOR shall advance to PERTAMINA before the
beginning of each annual Work Program a minimum amount
of seventy-five thousand U.S. dollars (U.S.$75,000.00) for the
purpose of enabling PERTAMINA to meet rupiah expenditures incurred pursuant to this paragraph c). If at any time
during the annual Work Program period the minimum amount
advanced under this paragraph c) has been fully expended,
separate additional advance payments as may be necessary to
provide for rupiah expenses estimated to be incurred by PER-
TAMINA during the balance of such annual Work Program
period will be made. If any amount advanced hereunder is not
expended by PERTAMINA by the end of an annual Work
Program period, such unexpended amount shall be credited
against the minimum amount to be advanced pursuant to this
paragraph c) for the succeeding annual Work Program period;

d) ensure that at all times during the term hereof sufficient
rupiah funds shall be available to cover the rupiah expenditure
necessary for the execution of the Work Program;

e) have title to all original data resulting from the Petroleum
Operations including but not limited to geological, geophysical, petrophysical, engineering, well logs and completion,
status reports and any other data as CONTRACTOR may
compile during the term hereof; provided, however, that all

such data shall not be disclosed to third parties without informing CONTRACTOR and giving CONTRACTOR the opportunity to discuss the disclosure of such data if CONTRACTOR so desires and further provided that CONTRACTOR may retain copies of such date;

f) to the extent that it does not interfere with CONTRACTOR'S performance of the Petroleum Operations use the equipment which becomes its property by virtue of this Contract solely for the Petroleum Operations envisaged under this Contract and if PERTAMINA wishes to use such equipment for any alternative purpose, then PERTAMINA shall first consult CONTRACTOR.

SECTION VI
RECOVERY OF OPERATING COSTS AND
HANDLING OF PRODUCTION

1. *Crude Oil*

1.1 CONTRACTOR is authorized by PERTAMINA and obligated to market all Crude Oil produced and saved from the Contract Area subject to the provisions hereinafter set forth.

1.2 CONTRACTOR will recover all Operating Costs out of an amount equal in value to a maximum of forty percent (40%) per annum of Crude Oil produced and saved hereunder and not used in Petroleum Operations, or such lesser percentage necessary to recover the Operating Costs as may from time to time be sufficient to accomplish this purpose, and except as provided in paragraphs d) and e) of subsection 1.1. Section VII, CONTRACTOR shall be entitled to take and receive and freely export such portion of Crude Oil. For purposes of determining the quantity of Crude Oil delivered to CONTRACTOR to recover said Operating Costs, the weighted average price of all Crude Oil produced and sold from the Contract Area during the Calendar Year will be used, excluding however deliveries made pursuant to subsection 1.2 paragraph p) of Section V. If, in any Contract Year, the Operating Costs exceed the value of the stated forty percent (40%), then the unrecovered excess shall be recovered in succeeding years.

1.3 Of the balance of the Crude Oil remaining PERTAMINA shall be entitled to take and receive seventy percent (70%) and CONTRACTOR shall be entitled to take and receive thirty percent (30%), provided however, that as to the quantity of daily production of Crude Oil which is in excess of fifty thousand (50,000) barrels per day up to and including one hundred thousand (100,000) barrels per day PERTAMINA shall be entitled to take and receive seventy-five percent (75%) and CONTRACTOR shall be entitled to take and receive twenty-five percent (25%), and provided further that as to the quantity of daily production of Crude Oil which is in excess of one hundred thousand (100,000) barrels per day, PERTAMINA shall be entitled to take and receive eighty percent (80%) and CONTRACTOR shall be entitled to take and receive twenty percent (20%).

1.4 Title to CONTRACTOR'S portion of Crude Oil as well as to such portion of Crude Oil exported and sold to recover Operating Costs shall pass to CONTRACTOR at the point of export.

1.5 CONTRACTOR will use its best reasonable efforts to market the Crude Oil to the extent markets are available. Either party shall be entitled to take and receive their respective portions in kind.

1.6 If PERTAMINA elects to take any part of its portion of Crude Oil in kind, it shall so advise CONTRACTOR in writing not less than ninety (90) days prior to the commencement of each semester of each Calendar Year specifying the quantity which it elects to take in kind, such notice to be effective for the ensuing semester of each Calendar Year (provided, however, that such election shall not interfere with the proper performance of any Crude Oil Sales Agreement for Petroleum produced within the Contract Area which CONTRACTOR has executed prior to the notice of such election). Failure to give such notice shall be conclusively deemed to evidence PERTAMINA'S election not to take in kind. Any sale of PERTAMINA'S portion of Crude Oil shall not be for a term of more than one Calendar Year without PERTAMINA'S consent.

1.7 The provisions regarding payment of CONTRACTOR'S Indonesian Income Tax shall be applied as follows:

a) CONTRACTOR shall be subject to the Indonesian Income Tax Laws and shall comply with the requirements of the law in particular with respect to filing returns, assessments of tax, and keeping and showing of books and records.

b) CONTRACTOR'S annual income for Indonesian Income Tax purposes shall be an amount equal to the sums received by CONTRACTOR out of the sale or other disposition of thirty percent (30%), twenty-five percent (25%), or twenty percent (20%), as the case may be of Crude Oil produced and saved and not used in Petroleum Operations after deduction of Operating Costs recovered under subsection 1.2 hereof plus an amount equal to CONTRACTOR'S Indonesian Income Tax.

....

e) As used herein Indonesian Income Tax shall be inclusive of all Income Tax payable to the Republic of Indonesia such as Company Tax, Income Tax, or Taxes based on income or profits including all dividend, withholding and other taxes imposed by the Government of Indonesia on the distribution of income or profits by CONTRACTOR.

2. *Natural Gas*

2.1 Any Natural Gas produced from the Contract Area to the extent not used in operations hereunder may be flared if the processing or utilization thereof is not economical. Such flaring shall be permitted to the extent that gas is not required to effectuate the maximum economic recovery of Petroleum by secondary recovery operations, including repressuring and recycling.

2.2 Should PERTAMINA and CONTRACTOR consider that the processing and utilization of Natural Gas is economical and choose to participate in the processing and utilization thereof, in addition to that used in secondary recovery operations, then the construction and installation of facilities for such processing and utilization shall be carried out pursuant to an approved Work Program. It is hereby agreed that, while it is the intention of the Parties to enter into further contractual arrangements to implement the foregoing, all costs and revenues derived from such processing, utilization and sale of Natural

Gas shall be treated on a basis equivalent to that provided for herein concerning Petroleum Operations and disposition of Crude Oil.

2.3 In the event, however, CONTRACTOR considers that the processing and utilization of Natural Gas is not economical, then PERTAMINA may choose to take and utilize such Natural Gas that would otherwise be flared, all costs of taking and handling to be for the sole account and risk of PERTAMINA.

SECTION VII
VALUATION OF CRUDE OIL

1.1 Crude Oil sold to third parties shall be valued as follows:

a) All Crude Oil taken by CONTRACTOR, including its share and the share for the recovery of Operating Costs, and sold to third parties shall be valued at the net realized price f.o.b. Indonesia received by CONTRACTOR for such Crude Oil.

b) All of PERTAMINA'S Crude Oil taken by CONTRACTOR and sold to third parties shall be valued at the net realized price f.o.b. Indonesia received by CONTRACTOR for such Crude Oil.

c) PERTAMINA shall be duly advised before the sales referred to in paragraphs a) and b) of this subsection are made, except for sale of CONTRACTOR'S share of Crude Oil.

d) Subject to any existing Crude Oil sales agreement, if a more favourable net realized price is available to PERTAMINA for the Crude Oil referred to in paragraphs a) and b) of this subsection, except CONTRACTOR'S share of Crude Oil, then PERTAMINA shall so advise CONTRACTOR in writing not less than ninety (90) days prior to the commencement of the deliveries under PERTAMINA'S proposed sales contract. Forty-five (45) days prior to the start of such deliveries CONTRACTOR shall notify PERTAMINA regarding CONTRACTOR'S intention to meet the more favorable net realized price in relation to the quantity and period of delivery concerned in said proposed sales contract. In the absence of such notice PERTAMINA shall market said Crude Oil.

e) PERTAMINA'S marketing of such Crude Oil as referred to in paragraph d) of this subsection shall continue until forty-five (45) days after PERTAMINA'S net realized price on said oil becomes less favorable. CONTRACTOR'S obligation to market said Crude Oil shall not apply until after PERTAMINA has given CONTRACTOR at least forty-five (45) days' advance notice of its desire to discontinue such sales. As long as PERTAMINA is marketing Crude Oil referred to above, it shall account to CONTRACTOR on the basis of the more favorable net realized price.

f) Without prejudice to any of the provisions of Section VI and Section VII CONTRACTOR may at its option transfer to PERTAMINA during any Calendar Year the right to market any Crude Oil which is in excess of CONTRACTOR'S normal and contractual requirements provided that the price is not less than the net realized price from the Contract Area. PERTAMINA'S request stating the quantity and expected loading date must be submitted in writing at least thirty (30) days prior to lifting said Crude Oil. Such lifting must not interfere with CONTRACTOR'S scheduled tanker movements. PERTAMINA shall account to CONTRACTOR in respect of any sale made by it hereunder.

1.2 Crude Oil sold to other than third parties shall be valued as follows:

a) by using the weighted average per unit price received by CONTRACTOR and PERTAMINA from sales to third parties excluding, however, commissions and brokerages paid in relation to such third party sales during the three (3) months preceding such sale adjusted as necessary for quality, grade and gravity;

b) if no such third party sales have been made during such period of time, then on the basis used to value Indonesian Crude Oil of similar quality, grade and gravity and taking into consideration any special circumstances with respect to sales of such Indonesian Crude Oil.

1.3 Third party sales referred to in this Section shall mean sales by CONTRACTOR to purchasers independent of CONTRAC-

TOR, that is purchasers with whom (at the time the sale is made) CONTRACTOR has no contractual interest involving directly or indirectly any joint interest.

1.4 Commissions or brokerages incurred in connection with sales to third parties, if any, shall not exceed the customary and prevailing rate.

1.5 During any given Calendar Year, the handling of production (i.e., the implementation of provisions of Section VI hereof) and the proceeds thereof shall be provisionally dealt with on the basis of the relevant Work Program and Budget based upon estimates of quantities of Crude Oil to be produced, of internal consumption in Indonesia, of marketing possibilities, of prices and other sale conditions as well as of any other relevant factor. Within thirty (30) days after the end of said given Year, adjustments and cash settlements between the Parties shall be made on the basis of the actual quantities, amounts and prices involved, in order to comply with the provisions of this Contract.

1.6 In the event the Petroleum Operations involve the segregation of Crude Oils of different quality and/or grade and if the Parties do not otherwise mutually agree:

a) any and all provisions of this Contract concerning valuation of Crude Oil shall separately apply to each segregated Crude Oil;

b) each Crude Oil produced and segregated in a given Year shall contribute to:

   (i) the 'total quantity' up to forty percent (40%) destined in such Year to the recovery of all Operating Costs pursuant to Section VI, subsection 1.2 hereof;

   (ii) The 'total quantity' of Crude Oil to which a Party is entitled in such Year pursuant to Section VI, subsection 1.3 hereof;

   (iii) the 'total quantity' of Crude Oil which CONTRACTOR agrees to sell and delivery in such Year for domestic consumption in Indonesia pursuant to paragraph p) of subsection 1.2 of Section V hereof, out of the share of Crude Oil to which it is entitled pursuant to Section VI, subsection 1.3, with quantities, each of which shall bear to the respective 'total quantity' (referred to in (i) or (ii) or (iii) above) the same proportion as the quantity of such Crude Oil produced and segregated in such given Year bears to the total quantity of Crude Oil produced in such Year from the Contract Area.

SECTION VIII
COMPENSATION AND PRODUCTION BONUS

1.1 CONTRACTOR shall pay to PERTAMINA as compensation for information now held by PERTAMINA the sum of Four Million U.S. Dollars (U.S.$4,000,000), and as a signature bonus the additional sum of Four Million U.S. Dollars (U.S. $4,000,000), after approval of this Contract by the Government of Indonesia in accordance with the provisions of applicable law. Such payments shall be due thirty (30) days after PER-TAMINA furnishes to CONTRACTOR an authenticated copy of such approval. On or before the date such payments become due PERTAMINA shall deliver to CONTRACTOR the geological and geophysical data held by PERTAMINA relating to the Contract Area.

1.2 CONTRACTOR shall pay to PERTAMINA the sum of One Million U.S. Dollars (U.S.$1,000,000) after daily production from the Contract Area averages Fifty Thousand (50,000) barrels per day for a period of one hundred twenty (120) consecutive days; and CONTRACTOR shall also pay to PER-TAMINA the sum of One Million U.S. Dollars (U.S. $1,000,000) after daily production from the Contract Area averages One Hundred Thousand (100,000) barrels per day for a period of one hundred twenty (120) consecutive days. Such payments shall be made within thirty (30) days following the last day of the one hundred twenty (120) days' period.

1.3 No part of such compensation and bonus payments shall be recovered by CONTRACTOR by inclusion in Operating Costs or otherwise.

SECTION IX
ADDITIONAL PAYMENTS

1.1 CONTRACTOR shall continue to make available to the domestic market in Indonesia Crude Oil as provided in para-

graph p) of subsection 1.2 of Section V hereof, and CON-
TRACTOR shall continue to receive Crude Oil allocated to
the recovery of Operating Costs in the manner and in the
amount provided in subsection 1.2 of Section VI hereof.

1.2 CONTRACTOR shall make additional payments to PER-
TAMINA which shall be computed solely on CONTRAC-
TOR'S share of Crude Oil after deducting from total produc-
tion of Crude Oil in the Contract Area (i) Crude Oil which
CONTRACTOR is entitled to receive as recovery of Operat-
ing Costs, (ii) Crude Oil made available to the domestic mar-
ket in Indonesia (assuming for this purpose that all the Parties
hereto supply shares to the domestic market on the basis
CONTRACTOR is required so to supply pursuant to para-
graph p) of subsection 1.2 of Section VI, and (iii) PER-
TAMINA'S share of Crude Oil pursuant to subsection 1.3 of
Section VI hereof.

1.3 The additional payments shall be calculated as follows:

a) With respect to any calendar quarter in which the
weighted average price per barrel (as determined hereunder for
purposes of recovery of Operating Costs, and hereinafter called
'realized price') of all Crude Oil produced and sold from the
Contract Area exceeds the Base Price as defined in paragraph
c) of this subsection 1.3. CONTRACTOR shall make the following additional payments to PERTAMINA: To the extent
that CONTRACTOR'S share of Crude Oil upon which additional payments are to be computed

(i) is within the thirty percent (30%) share which CON-
TRACTOR receives pursuant to subsection 1.3 of Section VI, the additional payment shall be fifty percent
(50%) of the amount by which the realized price of
Crude Oil exceeds the Base Price multiplied by the
number of barrels concerned;

(ii) is within the twenty-five percent (25%) share which
CONTRACTOR receives pursuant to subsection 1.3 of
Section VI, the additional payment shall be forty per-
cent (40%) of the amount by which the realized price of
Crude Oil exceeds the Base Price multiplied by the
number of barrels concerned;

(iii) is within the twenty percent (20%) share which

CONTRACTOR receives pursuant to subsection 1.3 of Section VI, the additional payment shall be twenty-five percent (25%) of the amount by which the realized price of Crude Oil exceeds the Base Price multiplied by the number of barrels concerned.

b) Examples of the computation of the additional payments are set out in Exhibit 'D'.

c) The Base Price shall be U.S.$5.00 per barrel, f.o.b. CONTRACTOR'S terminal, which shall be adjusted quarterly as provided next below. The adjustment of the Base Price shall be accomplished by multiplying U.S.$5.00 by a fraction whose numerator is the export price index for manufactured goods as reported in the United Nations Monthly Bulletin of Statistics for the last previous calendar quarter and whose denominator is such export price index for the first calendar quarter of 1974. In the event the amount of Base Price found by multiplying U.S.$5.00 by a fraction whose numerator is the weighted average value (as determined for purposes of recovering Operating Costs) per barrel of Crude Oil for the last previous calendar quarter and whose denominator is U.S.$10.80 is lower than the amount of Base Price arrived at by applying the formula in the previous sentence then such lower figure shall apply to find the Base Price for that calendar quarter. In the event the amount of Base Price found by multiplying U.S.$5.00 by a fraction whose numerator is the weighted average value per barrel of Crude Oil for the last previous calendar quarter and whose denominator is U.S.$10.80 is higher than the amount of Base Price arrived at by applying the formula in the second sentence of this paragraph, then again the lower amount shall apply to find the Base Price for that calendar quarter. (See Exhibit 'E' attached as to the application of this paragraph.) In the event the manufactured goods price index of the United Nations Monthly Bulletin of Statistics should no longer be published, then the Parties hereto or their successors shall mutually agree upon, the best available corresponding index designed to reflect the same economic circumstances, and that index shall be used in calculating adjustments under this paragraph. If such Parties should be unable to agree upon such a substitute index within six months of the time the index described above became unavailable, the President of the International

Chamber of Commerce (or, if the President is a national of the same country as any of the Parties hereto or their successors, then the Vice President of the International Chamber of Commerce) shall, upon request of any of the Parties hereto or their successors, designate the index to be used for this purpose.

d) When the pertinent final indices of the United Nations Bulletin of Statistics are not yet available at the time payment is due hereunder with respect to a calendar quarter, the Bulletin's initial preliminary index or, if that also is unavailable, the latest published index (whether preliminary index or final index) will be used, and appropriate adjustments will be made at the time the first payment becomes due after the final index for the calendar quarter concerned becomes available. When all the final indices and the other pertinent data to be used in making all the calculations provided herein become available with respect to an entire calendar year, the calculation will, on a weighted average basis, be adjusted to apply to such calendar year as a whole and appropriate adjustments in payment with respect to that calendar year will be made at the time of the next succeeding quarterly payment, following a procedure consistent with that provided in this contract.

e) It is understood that if for recovery of Operating Costs less than 40% of total Crude Oil sales is required, such lesser percentage will be taken out of the Base Price segment (i.e. total production multiplied by the Base Price) of such sales, up to 40% thereof, whereas any unrecovered balance will be deducted from the other price segment which is the difference between the Base Price Segment and the realized price as defined in paragraph a) of this subsection 1.3.

1.4 Contractor will pay the additional payment as adjusted on a quarterly basis in United States Dollars.

1.5 The value of any Crude Oil taken and received by CONTRACTOR pursuant to paragraph 1.3 of Section VI which is transferred by CONTRACTOR, without sale, into operations of CONTRACTOR other than those provided in this Contract, shall be the same as the value of Crude Oil taken for recovery of Operating Costs under subsection 1.2 of Section VI.

1.6 The provisions for additional payments as set forth herein are applicable only to Crude Oil in liquid form.

SECTION X
PAYMENT PROCEDURE

1.1 All payments which this Contract obligates CONTRACTOR to make to PERTAMINA shall be made in United States Dollars currency in New York, New York, U.S.A. at a bank to be designated by PERTAMINA and agreed upon by the Central Bank of Indonesia, or at CONTRACTOR'S election other currency acceptable to PERTAMINA, except that CONTRACTOR may make such payments in Indonesian Rupiahs to the extent that such currencies are realized as a result of the domestic sale of Crude Oil or Natural Gas or Petroleum products, if any. All such payments shall be translated at the rate applicable to all Petroleum Companies carrying on business in Indonesia.

1.2 All Payments due to CONTRACTOR shall be made by PERTAMINA in United States Dollars, or at PERTAMINA'S election other currencies acceptable to CONTRACTOR, at the CHASE MANHATTAN BANK at One Chase Manhattan Plaza, New York, New York, U.S.A. or at another bank to be designated by CONTRACTOR.

1.3 Any payments which PERTAMINA is required to make to CONTRACTOR and which CONTRACTOR is required to make to PERTAMINA pursuant to this Contract shall be made within thirty (30) days following the end of the month in which the obligation to make such payments occurs.

SECTION XI
TITLE TO EQUIPMENT

1.1 Equipment purchased by CONTRACTOR pursuant to the Work Program becomes the property of PERTAMINA when landed at the Indonesian ports of import and will be used in Petroleum Operations hereunder. Rental payments to PERTAMINA on equipment as defined in Exhibit 'C', Article II, subarticle 9, shall begin on the date of commencement of commercial production in the Contract Area and shall be at a rate commensurate with the useful life of the relevant asset, but not to exceed ten percent (10%) per annum, until the total of all such payments equals the purchase price. Should there be any inconsistency between the provisions of this Contract and the provisions of Exhibit 'C', then the provisions of subsection 1.2 of Section VI of this Contract shall prevail.

1.2 The provisions of subsection 1.1 of this Section XI shall not apply to leased equipment belonging to foreign third parties who perform services as a contractor which equipment may be freely exported from Indonesia.

SECTION XII
CONSULTATION AND ARBITRATION

1.1 Periodically, **PERTAMINA** and **CONTRACTOR** shall meet to discuss the conduct of the Petroleum Operations envisaged under this Contract and will make every effort to settle amicably any problem arising therefrom:

1.2 Disputes, if any, arising between **PERTAMINA** and **CON-TRACTOR** relating to this Contract or the interpretation and performance of any of the clauses of this Contract, and which cannot be settled amicably, shall be submitted to the decision of arbitration. **PERTAMINA** on the one hand and **CON-TRACTOR** on the other hand shall each appoint one arbitrator and so advise the other Party and these two arbitrators will appoint a third. If either Party fails to appoint an arbitrator within thirty (30) days after receipt of a written request to do so, such arbitrator shall, at the request of the other Party, if the Parties do not otherwise agree, be appointed by the President of the International Chamber of Commerce. If the first two arbitrators appointed as aforesaid fail to agree on a third within thirty (30) days following the appointment of the second arbitrator, the third arbitrator shall, if the Parties do not otherwise agree, be appointed, at the request of either Party, by the President of the International Chamber of Commerce. If an arbitrator fails or is unable to act, his successor will be appointed in the same manner as the arbitrator whom he succeeds.

1.3 The decision of a majority of the arbitrators shall be final and binding upon the Parties.

1.4 In the event the arbitrators are unable to reach a decision, the dispute shall be referred to Indonesian Courts of Law for settlement.

1.5 Except as provided in this Section, arbitration shall be conducted in accordance with the Rules of Arbitration of the International Chamber of Commerce.

SECTION XIII
EMPLOYMENT AND TRAINING OF
INDONESIAN PERSONNEL

1.1 CONTRACTOR agrees to employ qualified Indonesian personnel in its operations and after commercial production commences will undertake the schooling and training of Indonesian personnel for labor and staff positions including administrative and executive management positions. At such time, CONTRACTOR shall also consider with PERTAMINA a program of assistance for training of PERTAMINA'S personnel.

1.2 Costs and expenses of training Indonesian personnel for its own employment shall be included in Operating Costs. Costs and expenses of a program of training for PERTAMINA'S personnel shall be borne on a basis to be agreed by PERTAMINA and CONTRACTOR.

SECTION XIV
TERMINATION

1.1 This Contract cannot be terminated during the first three (3) years as from the Effective Date, except by provisions as stipulated in subsection 1.3 hereunder.

1.2 At any time following the end of the third year as from the Effective Date, if in the opinion of CONTRACTOR circumstances do not warrant continuation of the Petroleum Operations, CONTRACTOR may, by giving written notice to that effect to PERTAMINA and after consultation with PERTAMINA, relinquish its rights and be relieved of its obligations pursuant to this Contract, except such rights and obligations as related to the period prior to such relinquishment.

1.3 Without prejudice to the provisions stipulated in subsection 1.1 hereinabove, either Party shall be entitled to terminate this Contract in its entirety by ninety (90) days' written notice if a major breach of Contract is committed by the other Party, provided that conclusive evidence thereof is proved by arbitration or final court decision as stipulated in Section XII.

SECTION XV
BOOKS AND ACCOUNTS AND AUDITS

1. *Books and Accounts*

   PERTAMINA shall be responsible for keeping complete books and accounts with the assistance of CONTRACTOR reflecting all Operating Costs as well as monies received from the sale of Crude Oil, consistent with modern petroleum industry practices and proceedings as described in Exhibit 'C'. Until such time that commercial production commences, however, PERTAMINA delegates to CONTRACTOR its obligation to keep books and accounts.

2. *Audits*

   PERTAMINA or CONTRACTOR, as the case may be, shall have the right to inspect and audit the books and accounts relating to this Contract for any Calendar Year within the one (1) year period following the end of such Calendar Year. Any such audit will be satisfied within twelve (12) months after its commencement. Any exception must be made in writing within sixty (60) days following the end of such audit and failure to give such written exception within such time shall establish the correctness of the books and accounts.

APPENDIX C

Foreign Tax Credit, Indonesia; Oil and Gas Production-sharing Contracts

Amounts paid under certain oil and gas production sharing contracts to the Indonesian Government under its corporation and dividend tax laws and regulation with respect to these laws issued by the Minister of Finance effective January 1, 1978, are creditable income taxes for purposes of the foreign tax credit.

REV. RUL. 78–222[1]

Advice has been requested whether, under the circumstances described below, amounts paid to the Indonesian Government pursuant to the Indonesian Corporation Tax Ordinance of 1925, as amended by Law No. 8, 1970 (the Corporation Tax) and Law No. 12, 1959, as amended by Law No. 10, 1970 (the Dividend Tax) and in accordance with a Regulation of the Indonesian Minister of Finance, effective January 1, 1978 (The Regulation), are income taxes creditable under section 901(b) of the Internal Revenue Code of 1954.

P is a domestic corporation engaged with its affiliated companies, in the production, transportation, refining, and marketing of petroleum and petroleum products in the United States and abroad. S is a wholly owned United States subsidiary of P that joins with P in the filing of a Federal income tax return on a consolidated basis. S is engaged in income producing activities in Indonesia involving oil production. Under the Law of 1971 regarding State Oil and Natural Gas Mining Enterprises ('Pertamina Law'), all oil located in Indonesia is the property of the Indonesian Government. Pertamina, an Indonesian legal entity, wholly owned by the Government of Indonesia, was formed with the exclusive right to explore, develop, mine, and market Indonesian oil and gas.

Under Article 12 of the Pertamina Law, Pertamina may enter into production sharing contracts with non-government parties for the purpose of performing any of the above oil or gas related functions, provided such contracts conform with government regulations and are approved by the Indonesian President.

S presently holds several production sharing contracts that apply to different geographical regions within the jurisdiction of the Government of Indonesia. S entered into one such contract, a 30 year production sharing contract (Contract) with Pertamina in January 1975.

Under the Contract, S is required to: (1) start exploration within 5 months of signing the Contract; (2) spend at least $4,250x$ dollars on development over the first 8 years of the Contract; (3) supply a prescribed amount of oil produced and saved from the contract area to the Indonesian market at cost plus 20 cents per barrel (except that during the first 5 years of production from each new field in the contract area, the oil will be purchased by Indonesia at the weighted average of third party prices); (4) pay a signature bonus to Pertamina within 30 days of signing the Contract; (5) train and employ Indonesians; and (6) if specified conditions are satisfied, build and operate a refining facility in Indonesia. In addition, S must pay for all equipment used in its Indonesian operations and all expenses incurred in exploration, development, extraction, production, transportation, and marketing.

Pertamina and S divide the production that remains after S's recovery of its operating costs. To recover such costs, S must look solely to the extraction of oil and the income derived therefrom. Specifically, S recovers all operating costs in barrels of oil priced at market value prior to a division of oil between S and Pertamina. In general, all costs are recoverable. However, production bonuses and interest on money borrowed for petroleum operations are not allowable recovery costs under the contract. In addition, if in any year the amount of recoverable costs exceeds the total value of oil or gas produced and sold in that year, the excess may be carried forward to succeeding years, without limitation. Noncapital costs are recoverable in the year incurred and capital costs are recoverable over a period of years relating to the expected lives of recoverable reserves of oil. S is also entitled to receive additional cost recovery oil in an amount equal to 20 percent of the capital investment costs directly required for developing production facilities.

After S's recovery of such costs, the Contract provides that S is entitled to 34.0909 percent of the remaining production and Pertamina is entitled to the 65.9091 percent balance. From its share, S must then satisfy its obligation in accordance with the

Contract to supply oil to the Indonesian market.

S has the right to freely lift, dispose of, and export, the remainder of its share of production. The Government of Indonesia does not formally or informally require foreign-owned contractors to sell their share of exported Indonesian production at any official selling price, whether such sales are made to third parties or to affiliates. However, Indonesia does reserve the right to insure that amounts reported as gross income by contractors do in fact reflect the market value of production sold by them. 'Market value' will be determined on the basis of third party sales. For example, if a sale of Indonesian crude oil is made to a third party, the amount included in gross income will be the amount realized from that sale, assuming the absence of fraud. With respect to sales to affiliates, the market value shall be determined on the basis of comparable sales by Pertamina and contractors to third parties.

The Contract provides that *S* shall pay directly to the Indonesian Government the Corporation Tax and the Dividend Tax imposed on its income pursuant to the Indonesian Tax Law.

The Corporation Tax is a tax of general application imposed on entities engaged in business activities in Indonesia. The Corporation Tax is imposed on all taxable profits that are described therein as the net benefits derived. In the calculation of net benefits, gross benefits are reduced by the costs of acquisition, collection, and maintenance of such benefits. Thus the Corporation Tax is imposed on benefits or realized income and deductions are allowed for the significant expenses. Excess deductions from one activity may offset income from other activities.

The Regulation issued under the Indonesian Tax Law provides rules for the calculation of taxable profits in the case of oil or gas related income earned by contractors pursuant to production sharing contracts. The Regulation also specifies procedures for payment to the Corporation Tax and the Dividend Tax (Tax on Interest, Dividends, and Royalties) by such contractors and by certain technical assistance contractors.

Article 2 of the Regulation states that the basis of the application of the Corporation Tax and Dividend Tax to a contractor is taxable profit. Article 3 of the Regulation provides that the Corporation Tax is imposed at the rate of 45 percent of taxable profit and the Dividend Tax at the rate of 20 percent of taxable profit after deduction of the Corporation Tax.

In addition to being subject to such taxes on their production sharing contract income, contractors are also subject to such taxes on business income unrelated to production sharing contract income.

The Dividend Tax is imposed on 'proceeds' derived from corporations and associations subject to the Corporation Tax. The Dividend Tax provides that proceeds include annual profits received by partners or members, and payments on the annual profits, directly or indirectly, by a permanent establishment in Indonesia to enterprises abroad that have a right to it. The Dividend Tax of 20 percent is payable by the entity that earns the profits at the time they are available for payment. The Dividend Tax is imposed on such profits, however, regardless of whether or not the profits are actually remitted. Proceeds are considered available for payment as soon as the Corporation Tax is determinable.

Articles 4, 5, and 6 of the Regulation state that taxable profit shall equal gross income (the monetary value realized by a contractor for its share of products sold) reduced by the costs of obtaining, collecting, and maintaining such income, including interest on loans, and any bonuses paid by a contractor to Pertamina. Article 5 also provides that costs of surveys and intangible drilling costs may be deducted and that depreciation of fixed assets may be taken in accordance with procedures set forth in the Attachment to the Regulation. The Attachment to the Regulation specifies certain reasonable depreciation lives and recovery periods that can be shortened on the basis of the remaining proven reserves of crude oil.

Article 5 of the Regulation also provides that a contractor is required to treat all costs incurred with respect to a contract area prior to the commencement of production as preproduction costs. Preproduction costs that relate to capital equipment are recoverable through depreciation deductions in accordance with procedures set forth in the Attachment to the Regulation, beginning with the year in which such equipment is placed in service. Intangible drilling and development costs and all other costs not related to capital equipment are recoverable through amortization deductions beginning with the year in which production is begun in that contract area. The amortization deductions are taken over a period equal in length to that over which a contractor would depreciate production facilities. As provided in the Attachment to the Regulation, production facilities are depreciated over a 14 year period unless certain specific conditions exist, in which case the period is reduced to 7 years. The specific conditions are (1) proven reserves with 7 or fewer years of remaining production, or (2) situations in which the proven reserves divided by the estimated production of the succeeding year is 7 or less, or (3) new investment schemes related to production commencing in 1977 or later. The amount of the yearly amortization deduction for such preproduction costs is equal to the greater of: (1) the ratable portion of such costs for the taxable year, or (2) the net income for the contract area (after taking into account deductions for all costs including depreciation but before taking into account any amortization for the year). Any excess of such preproduction depreciation and amortization deductions may be subsequently used to offset the contractor's combined income from all contract areas and, if necessary, income from other activities.

Depreciation and amortization deductions are taken into account on an allowed or allowable basis. This means that such deductions will be considered taken when available to be taken and whether or not the full amount can be offset against income in that year.

Any costs that have not been deducted or utilized prior to the taxable year in which an area becomes worthless will be deducted in that year from the contractor's combined income from all areas. Costs incurred prior to the January 1, 1978, effective date of the Regulation, under contracts in effect on April 8, 1976, are, however, deductible last and only from income derived from the contract area in which the costs were incurred.

Article 5 of the Regulation further provides certain procedures for the computation of a contractor's taxable income with respect to several contract areas. A contractor that sustains losses in more than one contract area must apply such losses, in proportion to one another, against any income from another area or areas. That portion of any loss that remains is carried forward to subsequent years and is applied first against income from that contract area in which the loss was incurred. These provisions preserve the character of the losses incurred prior to January 1, 1978, under contracts in effect on April 8, 1976, and insure that such losses are only offset against income from the contract area in which such losses were incurred.

Article 7 of the Regulation requires that contractor to settle the

Corporation Tax and Dividend Tax by deposits to the account of the Indonesian Treasury with Bank Indonesia. In the event such deposits exceed tax liability as finally determined for the taxable year, the excess will be refunded.

Section 901(b) of the Code authorizes qualifying United States taxpayers to claim a foreign tax credit for the amount of any income tax paid or accrued during the taxable year to any foreign country or any possession of the United States. Section 1.901-2(b) of the Income Tax Regulations provides, in part, that the term 'foreign country' includes any foreign state or political subdivision thereof.

Rev. Rul. 78–61, 1978–8 I.R.B. 11 sets forth certain requirements for qualification of a payment as an income tax payment.

Rev. Rul. 76–215, 1976–1 C.B. 194, holds that no part of payments made pursuant to a production sharing contract described therein is a tax and no credit can be claimed for such payments under either section 901(b) or section 903 of the Code and no deduction can be taken under section 164(a) (3). Pursuant to the authority contained in section 7805(b), Rev. Rul. 76–215 is expressly inapplicable to amounts claimed as taxes paid or accrued to Indonesia for taxable years beginning before June 30, 1976, under production sharing contracts entered into prior to April 8, 1976. Pursuant to section 1035(c) of the Tax Reform Act of 1976 [Pub. L. 94–455, 1976–3 C.B. (Vol. 1) 1, 107], Congress extended this grace period to include taxable years ending on or before December 31, 1977.

Rev. Rul. 76–215 characterized the share of production of the Indonesian Government as a royalty and not a creditable income tax for the following reasons: (1) the production sharing contract provided the sole source of revenue for the Indonesian Government and the retained share of production was the Government's only compensation for the exhaustion of oil deposits to which the Government had title; (2) the recovery of signature and production bonuses and interest paid on borrowed money used for petroleum operations was not allowed and no recovery of annual operating costs in excess of 40 percent of the value of all barrels of oil produced and saved from the contract area during the year was allowed; (3) the income from each production sharing contract was computed separately from the income under other production sharing contracts held by the same party and a loss under one contract could not be offset against income earned under other contracts; and (4) the Indonesian Government was assured a share of production regardless of whether income had been realized.

Generally, in the absence of other factors that have contrary implications, payments to a foreign government owning the minerals in place extracted by a United States taxpayer will be treated as a creditable income tax if the characteristics described below are present.

First, in addition to payment of the income tax, the foreign government also requires payment of an appropriate royalty or other consideration for the property that is commensurate with the value of the concession. In the present case, the Indonesian Government receives payment of compensation for the concession in the form of a 65.9091 percent share of net production retained by Pertamina.

Second, the taxpayer's income tax liability cannot be discharged from property owned by the foreign government. In the present case, S's Corporation Tax and Dividend Tax liabilities are not discharged by the distribution to Pertamina of its 65.9091 percent share of net production or from any other property owned by Indonesia. Such taxes must be paid by S itself, from its own funds, to the Indonesian Treasury.

Third, the amount of income tax is calculated separately and independently of the amount of the royalty and of any other tax or charge imposed by the foreign government and satisfaction of such royalty or other charge is independent of the discharge of the foreign tax liability. In the present case, S's Corporation Tax and Dividend Tax payments and Indonesia's 65.9091 percent share of net production are mutually independent and separate in terms of amount and calculation. In addition, the distribution of that share of production to Indonesia does not discharge S's liability for the Corporation and Dividend Taxes.

Fourth, while the foreign tax base need not be identical or nearly identical to the United States tax base, the taxpayer, in computing the income subject to the foreign income tax, is allowed to deduct, without limitation, the significant expenses paid or incurred by the taxpayer. Reasonable limitations on the recovery of capital expenditures are acceptable. The only limitation on the deductibility of expenses relates to expenses incurred under the Indonesian Law in effect prior to the effective date of the Regulation, and, therefore, the Indonesian tax system herein considered allows a deduction of all significant expenses, incurred under current law, without limitation. In the present case, taxable income for purposes of the Corporation and Dividend Taxes is computed by deducting all significant expenses without limitation. While the deferral of the recovery of preproduction costs under the Regulation is not identical to the treatment of such costs for Federal income tax purposes, such deferral of recovery of preproduction costs under the Regulation is reasonable and does not preclude full recovery of costs.

Fifth, under the foreign law and in its actual administration, the income tax is imposed on the receipt of income by the taxpayer and such income is determined on the basis of arm's length amounts. Further, these receipts are actually realized in a manner consistent with United States income taxation principles. In the present case, payment of the Corporation and Dividend Taxes are only required if income is realized in any given taxable year in a manner consistent with United States tax principles. The calculation of taxable profits under the Regulation is determined on the basis of actual sales and is not dependent on an 'official' or 'posted' price; such a price does not function as part of the Indonesian tax system. See Rev. Rul. 78–63, 1978–8 I.R.B. 18.

Sixth, the foreign taxable income from extractive operations is computed on the basis of the taxpayer's entire extractive operations within the foreign country. The Regulation includes all amounts realized by S on the sale of its share of production in S's gross income. In addition, after January 1, 1978, any losses incurred in contract areas may be offset against income from other contract areas for purposes of calculation and payment of the Corporation Tax and Dividend Tax.

Seventh, the net taxable income or losses from extractive operations are combined with income or losses from other activities in applying the Corporation Tax and the Dividend Tax.

In addition to having the above characteristics, it is important to note that the Dividend Tax is a supplemental tax imposed on corporate profits at the time the Corporation Tax is determinable regardless of whether such profits are remitted. The Dividend Tax is imposed on the realization of earnings in the current year and not on the transfer of profits. A tax so imposed is regarded under United States standards to be a tax borne by the corporation earning the profits.

Accordingly, under the above facts, amounts paid pursuant to the Corporation Tax and Dividend Tax, in accordance with the Regulation, are creditable income taxes under section 901 of the Code.

1. Also released as News Release IR-1991, dated May 9, 1978, by the U.S. Internal Revenue Service.

APPENDIX D

TOTAL GROSS INVESTMENT OUTLAYS FOR OIL AND GAS
EXPLORATION, DEVELOPMENT AND PRODUCTION IN THE
ASIAN AREA RELATIVE TO TOTAL WELLS DRILLED, 1970–1977

Year	($ Million)	% Change	Total Drilled	% Change	Investment Outlay Per Well Drilled ($ Million)
1977	1,312	−13.2	700	− 4.5	1.874
1976	1,513	−14.5	733	− 1.6	2.064
1975	1,771	+38.9	745	− 8.0	2.377
1974	1,275	+78.3	810	+16.0	1.574
1973	715	+39.9	698	+30.4	1.024
1972	511	+30.6	535	− 7.5	0.955
1971	391	+41.6	579	+59.0	0.675
1970	276		364		0.758
Total	7,764		5,164		1.503

Source: Salomon Brothers. Well data, *Oil and Gas Journal*, 1977, estimated for wells
drilled.

As presented in E. Anthony Copp, 'Capital Sources versus Capital Demands in
Asian Petroleum Markets', paper presented at the 1978 Offshore Southeast Asia
Conference, Singapore, 21–24 February 1978.

APPENDIX E
JAVA SEA INVESTMENTS, 1972

	South-east Sumatra	Average	North-west Java	Average
Exploration				
Seismic: miles	17,100	—	12,000	—
cost	$ 2,982,000	$174/mi	$ 2,100,000	$175/mi
Wells: footage	114,300	—	90,200	—
cost	$17,647,000	$154/ft	$14,183,000	$157/ft
Delineation				
Wells: footage	28,300	—	49,100	—
cost	$ 4,259,000	$150/ft	$ 8,777,000	$179/ft
Production				
Platforms: number	2	—	7	—
cost	$ 1,130,000	$565,000	$ 3,150,000	$450,000
Wells: footage	36,100	—	86,600	—
cost	$ 4,583,000	$126.95/ft	$ 8,600,000	$99.31/ft
Producing facilities: cost	$ 4,680,000	—	$ 7,750,000	—
Pipelines & mooring: cost	$ 3,860,000	—	$ 3,600,000	—
Investment through 1971	$39,141,000	—	$48,160,000	—
Est. investment for 1972	$22,549,000	—	$29,316,000	—
Est. investment through 1972	$61,690,000	—	$77,476,000	—
Drilling costs per well, 1971				
Exploration	—	$415,000	—	NA
Development	—	$572,000	—	NA

Source: Developed from UNDP/CCOP *Technical Bulletin,* Vol. 11, p. 96.

APPENDIX F
AVERAGE EXPLORATORY WELL AND DEPTH COST
IN THE UNITED STATES, 1956–1975

Year	Average Well Depth (ft)	Average Well Cost (US$)	Cost per Foot (US$)
1956	4,022	50,200	12.48
1961	4,250	54,515	12.80
1966	4,465	68,386	15.32
1971	4,806	94,708	19.70
1972	4,932	106,424	21.58
1973	5,130	117,152	22.84
1974	4,660	131,000	28.10
1975	4,566	148,685*	32.56

Source: Extracted from *Petroleum Economist*, January 1977, p. 26.

*Estimated.

Bibliography

Economic Theory, Analysis and Policy

GENERAL THEORY AND METHODS

ALCHIAN, A. A. and ALLEN, W. R., *Exchange and Production: Theory in Use*, Belmont, California: Wadsworth, 1969.

ARROW, K. J., *Aspects of the Theory of Risk-Bearing*, Helsinki, 1965.

ARROW, K. J., KARLIN, S. and SCARF, H., *Studies in the Mathematical Theory of Inventory and Production*, Stanford, 1950.

BALCH, M. and WU, S. (eds.), *Essays on Economic Behaviour Under Uncertainty*, Amsterdam: North-Holland, 1974.

BAUMOL, W. J., *Business Behaviour, Value and Growth*, New York: MacMillan, 1959.

——, *Economic Theory and Operations Analysis*, Englewood-Cliffs, New Jersey: Prentice-Hall, 1965.

CHENERY, H. B., 'The Interdependence of Investment Decisions', in M. Abramovitz *et al.*, *The Allocation of Economic Resources*, Stanford, 1959.

COCHRANE, D. and ORCUTT, G. H., 'Application of Lease-Squares Regressions to Relationships Containing Auto-correlated Error Terms', *Journal of the American Statistical Association*, Vol. 44 (1949), pp. 32–61.

COHEN, B., *Multinational Firms and Asian Exports*, New Haven and London: Yale, 1975.

COHEN, K. J. and CYERT, R. M., *Theory of the Firm*, 2nd ed., New Delhi: Prentice-Hall, 1976.

CYERT, R. M. and MARCH, J. G., *A Behavioral Theory of the Firm*,

Englewood-Cliffs, New Jersey: Prentice-Hall, 1963.

DHRYMES, P. J., 'Alternative Asymptotic Tests of Significance and Related Aspects of 2 SLS and 3 SLS Estimates Parameters', *Review of Economic Studies*, Vol. 36 (1969), pp. 213–26.

DORFMAN, R., 'An Economic Interpretation of Optimal Control Theory', *American Economic Review*, Vol. 59 (1969), pp. 817–31.

ENCARNACION, J. JR., 'Constraints and the Firm's Utility Function', *Review of Economic Studies*, Vol. 31 (1964), pp. 113–19.

EZEKIEL, M. and FOX, K., *Methods of Correlation and Regression Analysis*, 3rd ed., New York: John Wiley, 1959.

FRIEDMAN, B. M., *Economic Stabilization Policy: Methods and Optimization*, Amsterdam: North-Holland, 1975.

FUROBOTN, E. and PEJOVICH, S., 'Property Rights and Economic Theory: A Survey of Recent Literature', *Journal of Economic Literature*, Vol. 10 (1972), pp. 1137–62.

——, *The Economics of Property Rights*, Cambridge, Mass.: Ballinger, 1974.

HALL, R. E. and JORGENSON, D. W., 'Tax Policy and Investment Behavior', *American Economic Review*, Vol. 57 (1967), pp. 391–414.

HIRSCHLEIFER, J., 'The Firm's Cost Function: A Successful Reconstruction', *The Journal of Business*, Vol. 35 (1962), pp. 235–55.

——, 'Investment Decision Under Uncertainty: Choice Theoretic Approaches', *Quarterly Journal of Economics*, Vol. 79 (1965), pp. 509–36.

HOERL, A. E. and KENNARD, R. W., 'Ridge Regression: Applications to Nonorthogonal Problems', *Technometrics*, February 1970, pp. 69–82.

——, 'Ridge Regression: Biased Estimation for Nonorthogonal Problems', *Technometrics*, February 1970, pp. 55–67.

HORST, T., 'American Taxation of a Multinational Firm', *American Economic Review*, Vol. 67 (1977), pp. 376–89.

JORGENSON, D. W., 'The Theory of Investment Behavior', in R. Ferber (ed.), *Determinants of Investment Behavior*, New York, 1967.

KMENTA, J., *Elements of Econometrics*, New York: MacMillan, 1971.

LIPSEY, R. G. and LANCASTER, K., 'The General Theory of

Second-Best', *Review of Economic Studies*, Vol. 24 (1956–7), pp. 11–32.

KNIGHT, F. H., *Risk, Uncertainty and Profit*, Boston: Houghton-Mifflin, 1921.

MITCHELL, C. C., 'Is the "Theory of the Firm" Misused in Current Land Economics Research?', *Land Economics*, Vol. 31 (1955), pp. 139–43.

MUTH, R., 'The Derived Demand for a Productive Factor and the Industry Supply Curve', *Oxford Economic Papers* (New Series), Vol. 16 (1964), pp. 221–37.

MYRDAL, G., *Economic Theory and Underdeveloped Regions*, London: University Paperbacks (no date—originally London: G. Duckworth, 1957).

NORDQUIST, G. L., 'The Breakup of the Maximization Principle', *Quarterly Review of Economics and Business*, Vol. 5 (1965), pp. 33–46.

PENROSE, E. T., *The Large International Firm in Developing Countries*, London: George Allen and Unwin, 1968.

ROTHENBERG, T. J. and SMITH, K. R., 'The Effect of Uncertainty on Resource Allocation', *Quarterly Journal of Economics*, Vol. 85 (1971), pp. 440–58.

RUMMEL, R. J. and HEENAN, D. A., 'How Multinationals Analyze Political Risk', *Harvard Business Review*, Vol. 56 (1978), pp. 67–76.

STIGUM, B. P., 'Entrepreneurial Choice over Time under Conditions of Uncertainty', *International Economic Review*, Vol. 10 (1969), pp. 426–42.

THEIL, H., *Economics and Information Theory*, Amsterdam: North-Holland, 1967.

——, 'A Note on Certainty Equivalence in Dynamic Planning', *Econometrica*, Vol. XXV (April 1957), pp. 346–9.

——, *Optimal Decision Rules for Government and Industry*, Amsterdam: North-Holland, 1964.

WILLIAMSON, O., 'Managerial Discretion and Business Behavior', *American Economic Review*, Vol. 53 (1963), pp. 1032–57.

WONG, R. E., 'Profit Maximization and a Dynamic Reconciliation', *American Economic Review*, Vol. 65 (1975), pp. 689–94.

GENERAL RESOURCE THEORY AND POLICY

BURT, O. R. and CUMMINGS, R. G., 'Production and Investment in Natural Resource Industries', *American Economic Review*, Vol. 60 (1970), pp. 576–90.

DASGUPTA, P. S. and HEAL, G. M., 'The Optimal Depletion of Exhaustible Resources', *Review of Economic Studies*, Symposium on Exhaustible Resources (1974).

DASGUPTA, P. S. and STIGLITZ, J. E., *Uncertainty and the Rate of Extraction Under Alternative Institutional Arrangements*, Technical Report No. 179, Institute for Mathematical Studies in the Social Sciences, Stanford, 1976.

DOUGLAS, A. J., 'Stochastic Returns and the Theory of the Firm', *American Economic Review*, Vol. 63 (1973), pp. 129–33.

GAFFNEY, M. (ed.), *Extractive Resources and Taxation*, Madison: University of Wisconsin Press, 1967.

GILBERT, R. J. and GOLDMAN, S. M., 'Potential Competition and the Monopoly Price of an Exhaustible Resource', Manuscript, presented at the annual meeting of the American Economic Association, 1977.

GOLDSMITH, O. S., 'Market Allocation of Exhaustive Resources', *Journal of Political Economy*, Vol. 82 (1974), pp. 1035–40.

GORDON, R. L., 'A Reinterpretation of the Pure Theory of Exhaustion', *Journal of Political Economy*, Vol. 75 (1967), pp. 274–86.

GRAY, L. C., 'Rent under the Assumption of Exhaustibility', *Quarterly Journal of Economics*, Vol. 28 (1914), pp. 466–89.

HERFINDAHL, O. C., 'Some Fundamentals of Mineral Economics', *Land Economics*, Vol. 31 (1955), pp. 131–8.

HOTELLING, H., 'The Economics of Exhaustible Resources', *Journal of Political Economy*, Vol. 39 (1931), pp. 137–75.

NORDHAUS, W., 'The Allocation of Energy Resources', *Brookings Papers on Economic Activities*, 1973, No. 3.

PEARCE, D. W. (ed.), *The Economics of Natural Resource Depletion*, New York: John Wiley, 1975.

SCOTT, A. D., 'The Theory of the Mine under Conditions of Certainty', in M. Gaffney (ed.), *Extractive Resources and Taxation*, Madison: University of Wisconsin, 1967.

SMITH, V., 'The Economics of Production from Natural Resources', *American Economic Review*, Vol. 58 (1968), pp. 409–31.

SOLOW, R. M., 'The Economics of Resources or the Resources of Economics', *American Economic Review*, Vol. 64 (1974), pp. 1–21.

STIGLITZ, J. E., 'Monopoly and the Rate of Extraction of Exhaustible Resources', *American Economic Review*, Vol. 66 (1976), pp. 655–61.

WEINSTEIN, M. and ZECKHAUSER, R. J., 'The Optimal Consumption of Depletable Natural Resources', *Quarterly Journal of Economics*, Vol. 89 (1975), pp. 371–92.

PETROLEUM AND ENERGY

ADELMAN, M. A., 'Economics of Exploration for Petroleum and Other Minerals', *Geoexploration*, Vol. 8 (1970).

——, 'Efficiency of Resource Use in Crude Petroleum', *Southern Economic Journal*, Vol. 31 (1964), pp. 101–20.

——, 'Long Run Cost Trends: Persian Gulf and United States', in Rocky Mountain Petroleum Economics Institute, *Balancing the Supply and Demand for Energy in the United States*, Denver: University of Denver, 1972.

——, *The World Petroleum Market*, Baltimore: Johns Hopkins University, 1972.

BRADLEY, P., 'Exploration Models and Petroleum Production Economics', in M. A. Adelman (ed.), *Alaskan Oil: Costs and Supply*, Praeger, 1971.

——, *The Economics of Crude Petroleum Production*, Amsterdam: North-Holland, 1967.

BRADLEY, P. and UHLER, R. S., 'A Stochastic Model for Determining the Economic Prospects of Petroleum Exploration over Large Regions', *Journal of the American Statistical Association*, Vol. 65 (1970), pp. 623–30.

DAVIDSON, P., 'Public Policy Problems of the Domestic Crude Oil Industry', *American Economic Review*, Vol. 53 (1963), pp. 85–108.

EPPLE, D. N., *Petroleum Discoveries and Government Policy*, Cambridge, Massachusetts: Ballinger, 1975.

ERICKSON, E. W., 'Economic Incentives, Industrial Structure, and the Supply of Crude Oil Discoveries in the United States, 1946–58/59', Unpublished manuscript.

ERICKSON, E. W. and SPANN, R. M., 'Supply Response in a Regulated Industry: the Case of Natural Gas', *Bell Journal of Economics and Management Science*, Vol. 2 (1971), pp. 94–121.

——, 'The U.S. Petroleum Industry', in E. Erickson and L. Waverman (eds.), *The Energy Question: An International Failure of Policy*, Toronto: University of Toronto, 1974.

FISHER, F. M., *Supply and Costs in the U.S. Petroleum Industry: Two Econometric Studies*, Washington, D.C.: Resources for the Future, 1964.

GORDON, R. L., 'A Reinterpretation of the Domestic Crude Oil Industry', *American Economic Review*, Vol. 53 (1963), pp. 85–108.

GRAYSON, C. J., *Decisions under Uncertainty: Drilling Decisions by Oil and Gas Operators*, Cambridge: Harvard University, 1960.

HUGHART, D., 'Informational Asymmetry, Bidding Strategies, and the Market of Offshore Petroleum Leases', *Journal of Political Economy*, Vol. 83 (1975), pp. 969–85.

JORGENSON, D. W. (ed.), *Econometric Studies of U.S. Energy Policy*, New York: Elsevier, 1976.

KALTER, R. J., TYNER, W. E. and STEVENS, T. H., *Atlantic Outer Continental Shelf Energy Resources: An Economic Analysis*, Ithaca, New York: Cornell, A.E. Res. 74–17, 1974.

KAUFFMAN, G. M., *Statistical Decisions and Related Techniques in Oil and Gas Exploration*, Englewood-Cliffs: Prentice-Hall, 1963.

KHAZZOOM, J. D., 'The FPC Staff's Econometric Model of Natural Gas Supply in the United States', *Bell Journal of Economics and Management Science*, Vol. 2 (1971), pp. 51–93.

KULLER, R. C. and CUMMINGS, R. G., 'An Economic Model of Production and Investment for Petroleum Reservoirs', *American Economic Review*, Vol. 64 (1974), pp. 66–79.

LELAND, H. E., NORGAARD, R. B. and PEARSON, S. R., 'An Economic Analysis of Alternative Outer Continental Shelf Petroleum Leasing Policies', Unpublished manuscript prepared for the Office of Energy R&D Policy, National Science Foundation, Washington, D. C., August 1974.

LOGUE, D. E., SWEENEY, R. and WILLETT, T., 'Optimal Leasing Policy for the Development of Outer Continental Shelf Hydrocarbon Resources', *Land Economics*, Vol. 51 (1975), pp. 191–207.

LOVEJOY, W. F. and HOMAN, P. T., *Economic Aspects of Oil Conservation Regulation*, Baltimore: Johns Hopkins University, 1967.

——, *Problems of Cost Analysis in the Petroleum Industry*, Dallas, Texas: Southern Methodist University, 1964.

MACAVOY, P. W. and PINDYCK, R. S., 'Alternative Regulatory Policies for Dealing with the Natural Gas Shortage', *Bell Journal of Economics and Management Science*, Vol. 4 (1973), pp. 454–98.

MANCKE, R. B., 'The Longrun Supply Curve of Crude Oil Produced in the United States', *Antitrust Bulletin*, Vol. 15 (1970), pp. 727–56.

MCCRAY, A. W., *Petroleum Evaluation and Economic Decisions*, Englewood-Cliffs, New Jersey: Prentice-Hall, 1975.

MEGILL, R. E., *Exploration Economics*, Tulsa: Petroleum Publishing Co., 1971.

MIKESELL, R. F., *Foreign Investment in the Petroleum and Mineral Industries*, Washington, D.C.: Resources for the Future, 1971.

NEWENDORP, P. D., *Decision Analysis for Petroleum Exploration*, Tulsa: Petroleum Publishing Co., 1975.

NORGAARD, R. B., 'Uncertainty, Competition, and Leasing Policy', Unpublished manuscript, Energy and Resources Group, University of California, Berkeley, 1977.

OEI, H. L., 'Petroleum Resource and Economic Development: A Comparative Study of Mexico and Indonesia', Ph.D. dissertation, University of Texas, Austin, 1964; Ann Arbor, Michigan: University Microfilms, Inc.

PINDYCK, R. S., 'The Regulatory Implications of Three Alternative Econometric Models of Natural Gas', *Bell Journal of Economics and Management Science*, Vol. 5 (1974), pp. 633–45.

SIDDAYAO, C. M., *The Off-shore Petroleum Resources of South-East Asia: Some Conflict Situations and Related Economic Considerations*, Kuala Lumpur: Oxford University Press, 1978.

STAUFFER, T. R., *The Economics of Taxation in the Eastern Hemisphere*, Cambridge, Massachusetts: A. D. Little, 1969.

SUDARSONO, B., 'Indonesian Energy Data and Projections', *Atom Indonesia*, Vol. 2 (1976), pp. 2–10.

SUKARNO, S., 'Monopolistic Competition in Indonesian Petroleum Industry', Unpublished Master's thesis, Naval Postgraduate

School, Monterey, 1973.

UNITED NATIONS, Economic and Social Commission for Asia and the Pacific, 'Energy Planning and Programming: Methodology of Planning—Assessment, Development, Demand, Supply (Storage, Where Relevant, and Distribution), Foreign Trade, Administration', Document NR/WGMEPP/2, 14 July 1978.

VAN MEURS, A. P. H., *Petroleum Economics and Offshore Mining Legislation*, Amsterdam: Elsevier, 1971.

WORLD BANK, *Energy and Petroleum in Non-OPEC Developing Countries, 1974–1980*, Staff Working Paper No. 229, February 1976.

Institutional Information

BOOKS, ARTICLES, REPORTS

ALBERS, J. P., CARTER, M. D., CLARK, A. L., COURY, A. B. and SCHWEINFURTH, S. P., *Summary Petroleum and Selected Mineral Statistics for 120 Countries, Including Offshore Areas*, Geological Survey Professional Paper 817, Washington, D.C.: U.S. Government Printing Office, 1973.

American Association of Petroleum Geologists, *Bulletin*, Annual issues on petroleum developments in the Far East, 1966–1977.

ARIEF, S., *The Indonesian Petroleum Industry*, Jakarta: Sritua Arief Associates, 1976.

——, *Financial Analysis of the Indonesian Petroleum Industry*, Jakarta: Sritua Arief Associates, 1977.

ARPS, J. J., 'Estimation of Primary Oil and Gas Reserves', in T. C. Frick (ed.), *Petroleum Production Handbook*, New York: McGraw-Hill, 1962.

ARPS, J. J., MORTADA, M. and SMITH, A. E., 'Relationship between Proved Reserves and Exploratory Effort', *Journal of Petroleum Technology* (June 1971), pp. 671–5.

AVERY, E. N., 'The Odds in Oil Exploration', UN, *Proceedings of the Third Symposium on the Development of Petroleum Resources of Asia and the Far East*, 1967, Vol. III.

BARTLETT, A. G. *et al.*, *Pertamina: Indonesian National Oil*, Jakarta: Amerisian, 1972.

BERG, R. R., CALHOUN, J. C. JR., and WHITING, R. L., 'Prognosis for Expanded U.S. Production of Crude Oil', *Science*, Vol. 184

(1974), pp. 331–6.

BOW VALLEY EXPLORATION (S) PTE. LTD., Press Release, 6 September 1978 (Calgary, Alberta, Canada).

BYBEE, R. W., 'Petroleum Exploration and Production on the Nation's Continental Shelves—Economic Potential and Risk', Paper presented at the Annual Meeting of the Marine Technology Society, 1970.

CAMERON, V. S., *Private Investments and International Transactions in Asian and South Pacific Countries*, New York: Matthew Bender, 1974.

CARLSON, S., *Indonesia's Oil*, Washington, D.C.: Georgetown University, Center for Strategic Studies, 1976.

CHANDLER, A. T., 'Mineral Exploration and Development: Some Basic Considerations/Trends in Government Management of Mineral Exploration and Development', UN, ESCAP Document No. E/ESCAP/NR.3/6., 21 July 1976.

CHANDLER, A. T. and TALMADGE, E. T. H., 'Summary of Petroleum Laws of Countries in the ECAFE Region', in United Nations, *Proceedings of the Seminar on Petroleum Legislation with Particular Reference to Offshore Operations*, ESCAP Mineral Resources Development Series, No. 40, 1973.

Chase Manhattan Bank, *Capital Investments of the World Petroleum Industry*, New York: annual.

——, *Financial Analysis of a Group of Petroleum Companies*, New York: annual.

CHUNG, S. K., GAN, A. S., LEONG, K. M., and KOH, C. H., 'Ten Years of Petroleum in Malaysia, 1966–1976', in UNDP/CCOP, *Technical Bulletin*, Vol. 11, Papers in commemoration of the Tenth Anniversary of CCOP, Bangkok, October 1977.

COPP, E. A., 'Capital Sources versus Capital Demands in Asian Petroleum Markets', Paper presented at the Offshore South East Asia Conference, Singapore, 21–24 February 1978.

DURKEE, F. F. and HATLEY, A. G., 'The Philippines: Is a Second Exploration Cycle Warranted?', *Oil and Gas Journal*, 18 January 1971, pp. 86–9.

FABRIKANT, R. (comp.), *The Indonesian Petroleum Industry: Miscellaneous Source Materials*, Field Report No. 4, Singapore: Institute of Southeast Asian Studies, March 1973.

——, *Legal Aspects of Production-Sharing Contracts in the Indonesian Petroleum Industry*, Field Report No. 3, Singapore: Institute of Southeast Asian Studies, March 1973.

FLOWER, A., 'World Oil Production', *Scientific American*, Vol. 238 (1978), pp. 42–9.

FREZON, S., *Summary of 1972 Oil and Gas for Onshore and Offshore Areas of 151 Countries*, U.S. Geological Survey Professional Paper 885, Washington, D.C.: U.S. Government Printing Office, 1974.

GAFFNEY, P. D., 'Petroleum Prospects in the Asia-Pacific Region', Pacific Basin Energy Conference, Singapore, 9–10 December 1974.

GAFFNEY, P. D. and MOYES, C. P., 'Competitive Legislation—the Key to Asia Pacific Petroleum Prospects', Paper presented at the Society of Petroleum Engineers' Session, Offshore Southeast Asia Conference, Singapore, 21–24 February 1978.

GAFFNEY, P. D., MOYES, C. P. and ALING, B., 'Economic Appraisal of the Potential Petroleum Resources of the Asian Pacific Region', Paper presented at the Offshore Southeast Asia Conference, SPE Session, Singapore, 19 February 1976.

GOLDSTONE, A., 'What was the Pertamina Crisis?', *Southeast Asian Affairs 1977*, Singapore: Institute of Southeast Asian Studies, 1977, by FEP International.

GROSSLING, B. F., *Latin America's Petroleum Prospects in the Energy Crisis*, U.S. Geological Survey Bulletin 1411, Washington, D.C.: U.S. Government Printing Office, 1975.

——, 'A Long-Range Outlook of World Petroleum Prospects', Prepared for the Subcommittee on Energy of the Joint Economic Committee, Congress of the United States, 2 March 1978.

——, 'The Petroleum Exploration Challenge with Respect to the Developing Nations', in R. Meyer (ed.), *The Future Supply of Nature-Made Petroleum and Gas*, Proceedings of UNITAR-IIASA Conference, Laxenburg, Austria, 5–16 July 1976, Pergamon Press, 1977.

HATLEY, A. G., 'Asia's Oil Prospects and Problems: An Overview of Petroleum Exploration Activity in East Asia', Paper presented at the SEAPEX Session, Offshore Southeast Asia Conference, Singapore, 21–24 February 1978.

——, 'The Nido Reef Oil Discovery in the Philippines—Its Significance', Paper presented at the ASEAN Council on Petroleum (ASCOPE) Conference, Jakarta, 11–13 October 1977.

——, 'Offshore Petroleum Exploration in East Asia—an Overview', Paper presented at SEAPEX Program, Offshore Southeast Asia Conference, Singapore, February 1976.

——, 'Oil and Gas Exploration and Development Programs—the Prerequisites', Paper presented at the Pacific Basin Energy Conference, Singapore, 9–10 December 1974.

ICHORD, R. F., *Energy Policies of the World: Indonesia*, Center for the Study of Marine Policy, University of Delaware, 1976.

——, 'Southeast Asia and the World Oil Crisis: 1973', *Southeast Asian Affairs, 1974*, Singapore: Institute of Southeast Asian Studies, 1974.

Indonesia, Department of Mining, *Indonesian Mining Yearbook*, 1974, 1975 and 1976 issues, Jakarta: Department of Mining.

——, Directorate General of Oil and Gas, *Indonesian Oil and Gas Data, 1977* (Bahan-bahan Informasi Minyak dan Gas Bumi Indonesia 1977), Jakarta.

——, Directorate General of Oil and Gas, *Indonesian Oil and Gas Industry of Indonesia, 1977*, monthly reports, Jakarta.

——, National Committee, World Energy Conference, *Hasil-hasil Lokakarya Energi: Perkiraan Kebutuhan Energi Indonesia, 1975–2000*, Papers, Jakarta, 12–13 May 1977.

——, National Committee, World Energy Conference, 1978 Workshop on Energy, 26–27 May 1978, Papers, Jakarta.

International Petroleum Encyclopedia, 1975 and 1977 issues.

LEICESTER, P., 'Risks and costs of oil exploration and development', in UN, *Proceedings of the Symposium on the Development of Petroleum Resources of Asia and the Far East*, 1959.

Malaysian Government, 'Energy Programming and Planning in Malaysia', Paper presented at the UN, ESCAP Working Group Meeting on Energy Planning and Programming, 15–21 August 1978, Bangkok, Document NR/WGMEPP/CRP.2, 15 August 1978.

MEYER, R. (ed.), *The Future Supply of Nature-Made Petroleum and Gas*,

Proceedings, UNITAR-IIASA Conference, Laxenburg, Austria, 5–16 July 1976, Pergamon Press, 1977.

MICHIE, M. S., 'The Search for Oil in Brunei', *Petroleum di-Brunei*, Brunei Shell Petroleum Company, April 1975.

MILLER, B. M., THOMSEN, H. L., DALTON, G. L., COURY, A. B., HENDRICKS, T. A., LENNARTZ, F. E., POWERS, R. B., SABLE, E. G. and VARNES, K. L., *Geological Estimates of the Undiscovered Recoverable Oil and Gas Resources in the United States*, Geological Survey Circular 725, Reston, Virginia: U.S. Geological Survey, 1975.

Mitre Corporation and the Malaysian Government, *Fifth International Symposium on Energy, Resources and the Environment*, Kuala Lumpur, 17–20 February 1975.

MONTEL, J., 'Necessary Conditions for the Development of Oil Search', in UN, *Proceedings of the Second Symposium on the Development of Petroleum Resources of Asia and the Far East*, 1963.

National Petroleum Council, *Enhanced Oil Recovery*, Washington, D.C., 1976.

NG, S. M., *The Oil System in Southeast Asia*, Field Report No. 8, Singapore: Institute of Southeast Asian Studies, 1974.

NORTHCUTT, E., *Summary of Mining and Petroleum Laws of the World*, Part 2: East Asia and the Pacific, Washington, D.C.: U.S. Bureau of Mines Circular 8514, 1971.

O'BRIEN, R., *South China Sea Oil: Two Problems of Ownership and Development*, Occasional Paper No. 47, Singapore: Institute of Southeast Asian Studies, 1977.

ODELL, P. R., *Oil and World Power: Background to the Oil Crisis*, London: Penguin, 1975.

Organization of the Petroleum Exporting Countries, *Selected Documents of the International Petroleum Industry, 1968*, Vienna.

Pertamina, *Indonesian Oil Statistics*, Jakarta.

Philippines, Republic of, *Pre-Investment Study on Power Industry Including Nuclear Power in Luzon, Philippines*, prepared for the United Nations by the Government of the Philippines, Manila, 1972.

——, Republic Act No. 387 (Petroleum Act of 1949).

——, Presidential Decrees No. 87 (21 December 1972), No. 782 (25 August 1975), No. 1352 (21 April 1978), No. 1442

(1978, specific date not available).

——, Petroleum Board, Rules and Regulations Governing Tax-Exempt Importations under Presidential Decrees No. 87 and 529.

——, Ministry of Energy, *Ten-Year Energy Development Program, 1978–1987*, Manila, 1978.

PRATT, W., 'Petroleum on the Continental Shelves', *Bulletin* of the American Association of Petroleum Geologists, Vol. 31 (1947), pp. 657–72.

RAJARETNAM, M., *Politics of Oil in the Philippines*, Field Report No. 6, Singapore: Institute of Southeast Asian Studies, 1973.

RAZALEIGH HAMZAH, TENGKU TAN SRI, 'Oil Industry and Its Impact on Malaysian National Development', *Indonesian Oil and Gas*, Vol. 2 (1976), pp. 3–6.

REIMANN, C. F., 'The Role of Private Risk Capital in the Discovery and Development of Petroleum Resources', in UN, *Proceedings of the Symposium on the Development of Petroleum Resources of Asia and the Far East*, 1959.

SCHANZ, J. J. JR., 'Oil and Gas Resources—Welcome to Uncertainty', *Resources*, No. 58 (March 1978), Washington, D. C.

SIDDAYAO, C. M., 'Looking for Oil in the Philippines', *Esso Eastern Review* (June 1966), New York.

——, 'Patterns in the Utilization of the Major Energy Resources of the United States', Report for the Ford Foundation, June 1974, partially reproduced in Ford Foundation, Energy Policy Project, *A Time to Choose*, Cambridge, Mass.: Ballinger, 1974.

Singapore, *Government Gazette*, Acts Supplement, No. 4, 10 March 1978.

——, Law Revision Commission, *The Statutes of the Republic of Singapore*, Vol. VI, Singapore: Government Printing Office, 1970.

SUDARSONO, B., SURJADI, A. J. and ILJAS, J., *Proyek Penelitian Perspektip Jangka Panjang Perekonomian Indonesia*, Report for LEKNAS-LIPI, Jakarta, 1977.

TSUMURA, A., 'General Factors to be Taken into Account in Estimating Capital Requirements for Exploration and Exploitation of Oil and Gas Resources', in United Nations, *Proceedings of the Fourth Symposium on the Development of Petroleum Resources of Asia and the Far East*, 1973.

Thailand, Department of Mineral Resources, *Ministerial Regulations Issued Under the Provisions of the Petroleum Act and the Petroleum Income Tax Act*, Bangkok, 1973.

——, *Petroleum Act (1971) and Amendment (1973)*, Bangkok, 1976.

——, *Petroleum Activities in Thailand*, Bangkok, 1976.

——, ESCAP Thai Delegation, Committee on Natural Resources.

——, *Energy in Thailand*, Document presented at UN, ESCAP, Committee on Natural Resources, Fifth Session, 31 October–6 November 1978, Bangkok.

——, *Progress Report in Energy Development in Thailand*, Presented at ECAFE, Committee on Natural Resources, second session, 14–20 October 1975, Bangkok.

United Nations, *World Energy Supplies, 1950–1974*, Statistical Series J, No. 19, New York, 1976.

——, *World Energy Supplies, 1971–1975*, Statistical Series J, No. 20, New York, 1977.

United Nations, ECAFE, *Case Histories of Oil and Gas Fields in Asia and the Far East*, Mineral Resources Development Series No. 20, New York, 1963; Mineral Resources Development Series No. 29, New York, 1967; Mineral Resources Development Series No. 37, New York, 1971.

——, 'Energy Resources in the Region—Progress in Energy Development', Document NR/WGMEPP/1, 6 July 1978, Bangkok.

——, *Mineral Resources of the Lower Mekong Basin and Adjacent Areas of Khmer Republic, Laos, Thailand, and Republic of Viet-Nam*, Mineral Resources Development Series No. 39, 1972 (?).

——, *Proceedings of the Symposium on the Development of Resources of Asia and the Far East*, Mineral Resources Development Series No. 10, Bangkok, 1959.

——, *Proceedings of the Second Symposium on the Development of Petroleum Resources of Asia and the Far East*, Mineral Resources Development Series No. 18, Vols. I, II, and III, New York, 1963.

——, *Proceedings of the Third Symposium on the Development of Petroleum Resources of Asia and the Far East*, Mineral Resources Development Series No. 26, Vols. I, II, and III, New York, 1967.

——, *Proceedings of the Fourth Symposium on the Development of Petroleum*

Resources of Asia and the Far East, Mineral Resources Development Series No. 43, Vols. I, II, and III, New York, 1973.

——, *Proceedings of the Seminar on Petroleum Legislation with Particular Reference to Offshore Operations*, Mineral Resources Development Series No. 40, New York, 1973.

United Nations, ESCAP, *Proceedings of the Twelfth Session of the Sub-Committee on Energy Resources and Electric Power*, New York, 1974.

United Nations, CCOP, *The Offshore Hydrocarbon Potential of East Asia: A Review of Investigations, 1966–1973*, Bangkok, 1974.

——, *The Offshore Hydrocarbon Potential of East Asia: A Decade of Investigations (1966–1976)*, Bangkok, 1976.

——, Annual Reports.

——, *Technical Bulletin*, Vol. 11, Bangkok, 1977.

United States, Bureau of Mines, *Mineral Industry Surveys*, 'World Crude Oil Production, Annual', 'World Natural Gas, Annual'.

——, *Offshore Petroleum Studies*, B. M. Circular 8557, Washington, D.C., 1972.

United States, Council on Environmental Quality, *OCS Oil and Gas—An Environmental Assessment*, Vols. 1–5, Washington, D.C.: U.S. Government Printing Office, 1974.

United States, Federal Energy Administration, *The Relationship of Oil Companies and Foreign Governments*, Washington, D.C., 1975.

United States, Embassy, Jakarta, *Indonesia's Petroleum Sector*, 1977 and 1978 issues.

United States, Internal Revenue Service, *Rev. Ruling 78–222*, Released as News Release IR-1991 dated 9 May 1978.

WEAVER, L. K. *et al.*, *Composition of the Offshore U.S. Petroleum Industry and Estimated Costs of Producing Petroleum in the Gulf of Mexico*, Washington, D.C.: U.S. Department of the Interior, 1972.

Workshop on Alternative Energy Strategies (WAES), *Energy: Global Prospects, 1985–2000*, Report sponsored by the Massachusetts Institute of Technology, McGraw-Hill, 1977.

World Energy Conference, *Survey of Energy Resources*, London, 1974.

PERIODICALS

Asia Research Bulletin, Vol. 8 (January 1979), 'Energy and Mineral Resources'.

Asian Wall Street Journal, 7 February 1977, pp. 1, 10, 'Indonesia plans new incentives for oil firms'.

——, 21 February 1977, p. 3, 'Indonesia's new exploration incentives disappoint most foreign oil companies'.

——, 11 March 1977, pp. 1, 5, 'Indonesia might allow foreign participation in state's oil fields'.

——, 23 March 1977, p. 10, 'Indonesia to cut oil exports'.

——, 14 April 1977, p. 3, 'Indonesia opens Pertamina oil fields to participation by foreign companies'.

——, 19 May 1977, p. 3, 'Pertamina and Caltex sign pact to explore offshore Irian Jaya'.

——, 20 May 1977, p. 3, 'Indonesia gives oil firms breaks to spur drilling'.

——, 7 June 1977, p. 3, 'Vietnam enacts foreign investment code with eased terms on exports, ownership'.

——, 6 July 1977, p. 3, 'Pertamina seen successful in bid for exploration'.

——, 24 October 1977, p. 3, 'Pertamina signs pacts for areas previously reserved for state'.

——, 19 April 1978, p. 3, 'Vietnam gives Italy rights to seek oil in its coastal waters'.

——, 5 May 1978, p. 3, 'Philippines plans Nido oil production in 2nd period of '79'.

——, 26 May 1978, p. 3, 'Indonesia ready to let first contracts in $1 billion Sumatran coal, power plan'.

——, 15 June 1978, p. 2, 'Waste gas at oil fields may be fuel in Malaysia'.

——, 20 September 1978, pp. 1, 12, 'Indonesian officials growing impatient on LNG sales to U.S.'.

——, 24 October 1978, pp. 1, 12, 'Thailand still faces hurdles in pricing Texas Pacific gas'.

——, 8 December 1978, pp. 1, 11, 'Major oil find considered likely for Philippines'.

——, 19 December 1978, p. 3, 'Indonesia signs pact with Italy to set up nuclear power plant'.

——, 14 February 1979, p. 2, 'Canada firms search for oil off Vietnam'.

Far Eastern Economic Review, 13 August 1976, pp. 32–3, 'Jakarta's bitter oil men weigh up their future'.

——, 8 July 1977, pp. 36–7, 'Petronas digs in its heels'.

Indonesian Times (Jakarta), 26 May 1978, p. 1, 'Nat. Coordinating Board on energy to be set up'.

Oil and Gas Journal, Annual December issues on petroleum data.

——, 20 June 1977, pp. 34–5, 'U.S. wildcat-success rate hits 16.88%'.

——, 30 January 1978, p. 144, 'Giant fields still yield most U.S. oil'.

——, 31 July 1978, p. 102, 'Viet-Nam claims more Hanoi trough discoveries'.

——, 28 August 1978, pp. 37–8, 'Esso: Far East oil potential limited'.

The Petroleum Economist, January 1975, pp. 21–4, 'From concessions to contracts—II: End of an era'.

——, October 1976.

——, May 1978, p. 212, 'Agip secures exploration permits in Vietnam'.

——, June 1978, pp. 248–9, 'Brunei: The "Gulf State" of the Far East'.

——, November 1978, p. 487, 'Indonesia'.

——, December 1978, pp. 505–6, 'Burma—call for exploration bids'.

——, February 1979, pp. 62–5, 'Vietnam—problems loom for renewed search', and p. 83, 'Selected Crude Oil Prices'.

——, April 1979, p. 174, 'Far East'.

——, May 1979, p. 214, 'Far East'.

——, January 1980, pp. 31–2, 'Rush for exploration contracts'.

Petroleum News, January issues, 1974–9.

——, April 1973, pp. 28–9, 'Vietnam'.

——, December 1976, p. 13, 'Paying one's taxes'.

——, June 1977, p. 16, 'Hydropower in Peninsular Southeast Asia'.

——, July 1978, p. 18, 'Malaysia cuts waste, ups production'.

——, August 1978, pp. 8, 10, 'Vietnam wants American oil firms'.

——, September 1978, p. 3, 'Exploration highlights (Vietnam)'.

——, December 1978, supplement.

Petromin Asia, May 1978, 'Offshore Vietnam: Deminex first to sign production contract'.

——, June 1978, p. 10, 'Norway steps up aid to Vietnam for oil exploration'.

Straits Times (Singapore), 17 March 1977, p. 4, 'Jakarta's 50–50 offer for oil search in remote areas'.

——, 3 September 1977, p. 8, 'Coal to replace oil study by PUB'.

——, 22 October 1977, p. 6, 'Switch likely to coal fuel in new power station'.

——, 14 December 1977, p. 6, '49 firms "yes" to new oil pact'.

Index